WELDING
COMPLETE
2ND EDITION

TECHNIQUES, PROJECT PLANS & INSTRUCTIONS

MICHAEL A. REESER

COOL
SPRINGS
PRESS

Quarto is the authority on a wide range of topics.

Quarto educates, entertains and enriches the lives of our readers—enthusiasts and lovers of hands-on living.

www.quartoknows.com

First published in 2009 by Creative Publishing international, an imprint of The Quarto Group, 401 Second Avenue North, Suite 310, Minneapolis, MN 55401 USA. This edition published 2017 by Cool Springs Press. Telephone: (612) 344-8100 Fax: (612) 344-8692

quartoknows.com
Visit our blogs at quartoknows.com

Cool Springs Press titles are also available at discount for retail, wholesale, promotional, and bulk purchase. For details contact the Special Sales Manager by email at specialsales@ quarto.com or by mail at The Quarto Group, 401 Second Avenue North, Suite 310, Minneapolis, MN 55401 USA.

10 9 8 7 6 5 4 3 2 1

ISBN: 978-1-59186-691-6

Library of Congress Cataloging-in-Publication Data

Names: Cool Springs Press, author. | Creative Publishing International, author.
Title: Welding complete.
Description: 2nd edition. | Minneapolis : Cool Springs Press, 2017. | "First published in 2009 by Creative Publishing international, an imprint of Quarto Publishing Group USA Inc."--Verso title page. | Includes bibliographical references and index.
Identifiers: LCCN 2016059357 | ISBN 9781591866916 (plc)
Subjects: LCSH: Welding.
Classification: LCC TS227 .W357 2017 | DDC 739/.14--dc23
LC record available at https://lccn.loc.gov/2016059357

Acquiring Editor: Todd Berger
Project Manager: Alyssa Bluhm
Art Director: Brad Springer
Book Designer: Simon Larkin
Layout: Rebecca Pagel
Photography: rau+barber
Edition Editor: Mark Schwendeman

Printed in China

CONTENTS

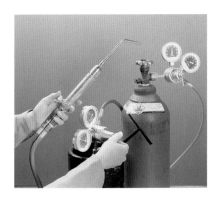

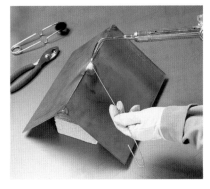

INTRODUCTION

Welding is a practical skill that is challenging, rewarding, and also great fun. We encounter welded items throughout our daily lives and activities—they are practically infinite. The appliances in our homes, the railing on our porches or stairs, our automobiles, the bridges and infrastructure we drive on, the structure of the buildings we work in, and even our computers. Welding makes everyday tasks more manageable. Welding can also bring joy to a craftsman making items for use in the shop, home décor, and outdoor living and décor.

In this book, we will thoroughly discuss the basics of welding and fabrication: safety, equipment, material selection, tools, and the various welding and cutting processes. With a firm understanding of the basics, applying safety and common sense, we will then move on and apply your newfound knowledge to specific projects in the book, complete with how-to instructions, tool lists, material lists, blueprints and illustrated pictorial plans, and photos. Developing your new welding and fabrication skills will provide hours of satisfaction as you create items and furnishings for your home, shop, and outdoor living space.

The shop section includes plans and instructions for making practical and useful items specific for your working space that will provide years of useful service. These tools include a rolling welding curtain, welding and cutting table, welding cart with cylinder rack, and an expandable welding table.

The home décor section includes plans and instructions for commonly used items, such as coat hooks, tables, and shelves, as well as decorative items, table candelabrum, and decorative wall fixtures.

The outdoor section includes plans and instructions for projects suitable for all hardworking DIYers, including a yard trailer and truck rack. Also included are plans and instructions for outdoor living space items, such as railings, a garden/yard gate, arbor, and firepit.

It is important as a welder to understand and use the correct terminology when discussing welding and the related processes. This will be useful in ordering equipment, and in selecting the proper filler metal and materials. The *weldment* (the parts to be joined) is referred to as the base metal. Additional metal, called the *filler metal*, is then added to the molten base metal to form a molten puddle that will solidify into a new metal—this is now known as the heat affected zone (HAZ). This is the area of focus for the welder. To maintain the proper fusion and strength of the newly formed weld, it is very important for the base metal and filler metal to have the same composition. The processes for joining metal without fusion are soldering, brazing, and braze welding. These processes can be used to join either similar or dissimilar metals.

The strength and quality of any weld is dependent on many factors. To achieve the proper amount of penetration, or weld fusion, it is critical that the heat input is controlled through an understanding of how to set, adjust, and maintain the following variables:

• Base metal selection
• Filler metal selection
• Proper heat input for a selected material thickness
• Work angle
• Travel angle
• Arc length
• Travel speed
• Aim
• Electrode manipulation
• Joint design

Welding is about managing heat input—whether it is from a fuel source such as oxyacetylene welding (OAW), or from electricity in an arc welding system, such as shielded metal arc welding (SMAW), gas metal arc welding (GMAW), flux core arc welding (FCAW), or gas tungsten arc welding (GTAW). Each process has its own

SAFETY

Welding can be a dangerous activity. Failure to follow safety procedures may result in serious injury or death. This book will provide useful instruction, but we cannot anticipate all working conditions presented while performing welding and cutting or the characteristics of your material and tools. Safety is applying good judgment and common sense—you should use caution and care when following the procedures described in the book. ALWAYS consider your own skill level and the Safety Notices associated with each tool, and use them properly; STOP and consult the owner's manual or manufacturer for any questions. The publisher, author, and technical director cannot assume responsibility for any damage to property or injury to persons as a result of misuse of the information provided.

Oxyacetylene welding.

Shielded metal arc welding (SMAW).

Gas metal arc welding (GMAW).

advantages and disadvantages. Once a welding process is selected, the goal is to join the selected parts to form a useful tool or item in a permanent manner. Learning to manage the heat input allows the welder to control the molten metal *puddle* or the HAZ, thus allowing the base metal, original parts, and filler metal to flow and fuse into a new coalesced area, joining the parts into a weldment.

OAW and Oxyacetylene cutting (OAC) use acetylene for the fuel source to produce flames to generate heat to melt the base and filler metal for welding or for cutting of ferrous metals. Acetylene is the fuel of choice for welding, because it is the only fuel source that will generate a truly "neutral" flame. When alternative fuel sources such as propane or natural gas are used, the process is then referred to as Oxy-fuel welding (OFW) and Oxy-fuel cutting (OFC). These fuels are often selected for cutting purposes because of their lower cost and ease of access.

While performing OAW, the formation of the puddle is easier to see, as it is a slower process. The welder is watching for a color change of the base metal as it approaches the melting temperature. As the temperature increases, the color reaches a reddish color and appears glossy as it starts to melt (*wet out*). This wetting action allows the melting base metal to flow or join with the filler metal that is added to the joint to form the new metal in the HAZ. This forms a seamless molten area that will solidify into new metal.

The arc welding (AW) processes—SMAW, GMAW, FCAW, GTAW, and PAC—all use electrical current to produce an arc to generate the heat necessary to melt the base and filler metal to form the weld. With the AW process, the arc forms the puddle quickly and may be difficult to see without the proper lens shade, due to the intense light created by the arc. This is why wearing the proper shade of filter lens is important.

Penetration of the weld is also a critical heat-dependent factor. A strong weld penetrates all the way through the base metal. To ensure a successful, completely fused weld, the filler metal size, heat input (welding current), and base metal thickness must be matched to travel speed, travel angle, and arc length. It is easy to achieve an appropriately shaped weld profile that has not penetrated the base metal at all and merely sits on the surface. This is known as a "cold" weld and is associated with insufficient current. An opposite problem is "burn through"—where the current is set too high or the arc length is too long and overheats the base metal, making the puddle difficult to maintain and eventually burning through the base metal, leaving a hole that can be difficult to correct or repair.

Distortion caused by heat applied during all welding and cutting operations, whether by flame or electrical arc, is an unwanted by-product the welder must learn to identify. Pre-welding setup, welding sequence, and post-weld heat management are key to accurate dimensions. The welding or cutting process selected can produce flame or arc with a temperature of up to 10,000 degrees Fahrenheit. As we apply the arc (heat or current) to melt the base metal and filler metal to form the new material in the HAZ, the metal expands; as the metal cools, it contracts, causing potential stresses to form in the metal and the weld joint. If the expansion and contraction of the metal are not considered during the fit-up, the welding process can cause parts to move out of alignment or ultimately fail. Although the welder can clamp and/or tack weld parts together, he or she still needs to consider the stress in the HAZ and/or surrounding area. Clamping or improperly located tacks can alter where the stress concentrates, potentially causing premature failure of the weldment.

It is also imperative that the welding process is properly matched for each type and thickness of base metal. For example, SMAW is typically best suited for welding on material 3/16 inch or thicker, due to the heat input of the arc and difficulty in maintaining the arc on thinner material. GMAW, on the other hand, is well suited for thinner material or sheet metal.

As the arc is developing the HAZ or welding puddle, it is critical that atmospheric air—mainly oxygen and nitrogen—is kept away from the puddle while it is molten and as it cools or solidifies. Oxygen and nitrogen that contaminate the puddle as it cools will produce a very weak and brittle material. The welding process selected provides a protective area around the puddle to shield it as it cools or solidifies. In OFW, the properly adjusted neutral flame burns off ambient oxygen in a small zone around the weld puddle. GMAW and GTAW processes utilize an externally applied inert shielding gas to protect the puddle from atmospheric air as it solidifies; this is accomplished by attaching an external high-pressure cylinder. The most common shielding gas utilized is argon, chosen because it is denser than air and settles around the weld puddle, and because it is inert and does not react with the atmospheric air or the weld puddle. SMAW and FCAW processes use fluxes (chemical compounds) added in or on the filler metal. When these fluxes burn or melt, they produce shielding gases and form a protective coating (slag), both of which protect the weld area until it has solidified.

Welding can be difficult and takes years to master, but with basic knowledge and lots of practice, it is possible to make many useful and decorative items. If you wish to move beyond the projects outlined in this book, talk with a more experienced welder and have him or her evaluate some of your practice welds. Remember the safety of others is involved when you choose to make a utility trailer or spiral staircase. Take the time and make the effort to ensure that any project you make is safe.

This book is intended as a reference for people who have had some exposure to welding and who can follow the steps and safety precautions outlined in each project. It is not intended to teach welding to someone who has never handled welding equipment. If you wish to further your welding experience and improve your skill set, many community colleges, technical colleges, and art centers offer welding classes. Such classes are an ideal way to learn the basics of welding, the proper techniques associated with each process, and specific welding safety.

Gas tungsten arc welding (GTAW).
Chris Alleaume/ Alamy Stock Photo

Plasma arc cutting (PAC).

Oxyacetylene cutting (OAC).

BASICS

The costs involved in setting up a well-equipped home welding shop are comparable to setting up a well-equipped woodworking shop. A knowledgeable welder with the proper equipment has a great advantage, because the projects and repairs he or she can accomplish are too numerous to count. The availability of numerous inexpensive materials is also key to being creative. From basic repairs, to building your own shop tools and work stations, to creating items you can sell—the possibilities with welding are only limited by your imagination and experience. Always striving to learn and improve your craft lets you experience the "wow" factor when others see and admire your work.

SAFETY

Safety is very important when welding. The potential dangers inherent to welding and cutting processes are numerous. Failure to follow or overlook manufacturer's Material Data Safety Sheets (MSDS), recommended practices, and guidelines can result in injury or death. Welders may encounter such hazards as electrical shock, exposure to fumes and gases, fire, explosion, arc radiation burns, and cuts and abrasions. Proper preparation for all welding and cutting operations is critical for your personal safety and the safety of those around you. Always familiarize yourself with ALL safety information pertaining to a piece of welding or cutting equipment, and apply the information outlined in American National Standards Institute (ANSI) publication Z49.1, Safety in Welding, Cutting, and Allied Processes. A free copy of this standard can be downloaded from the American Welding Society website, www.aws.org. Important: when in doubt or when you have a question, consult the operator's manual or manufacturer's website, and remember safety is your first responsibility. Follow all safety rules and never take shortcuts.

FUMES. Welding produces potentially hazardous fumes and gases. Always keep your face out of the weld plume, and avoid breathing concentrations of those fumes. Welding indoors requires special precautions. When and what type of ventilation to select is the first and most important question to ask when welding indoors. Typically, any area with a ceiling lower than 16 feet or a width a total area of less than 10,000 square feet will require forced ventilation. A fan, exhaust hood, or fume extractor may be used. If adequate ventilation cannot be provided, an OSHA-approved respirator or particle mask may be needed. Always follow the fitting guidelines when purchasing a respirator. If you may be pregnant or plan to become pregnant, consult your physician. Never cut or weld on a container or material if you are unsure about its prior use or the type of coatings that it may have applied to the surface. Cutting or welding on containers or materials that contain toxic chemicals can be potentially deadly.

BURNS. Hot sparks, flying slag, and molten metal can cause severe burns in the welding shop. Always protect any exposed skin or body parts with the proper personal protective equipment (PPE). The most common materials to use are natural fibers—leather, wool, or cotton. **Never** wear synthetic materials, as they can melt to skin and cause more severe damage and burns. Wear leather boots with pants that fall over the top of the boots. Do not roll up or cuff pant legs or shirt sleeves, as they may trap sparks or slag. Make sure pants are free from frayed

Welding safety includes protecting yourself with safety gear, following manufacturer's instructions, and being aware of the surroundings in which you are welding.

edges and holes, as they are also a fire hazard. Wear a welding cap and keep hair tucked away.

ARC BURN. Welding arcs produce ultraviolet and infrared light. Both of these can damage your eyes permanently, burn your skin, and potentially lead to skin cancer. Always protect your face and eyes by wearing a welding helmet with a shaded lens of the appropriate number for the selected process. Wear a long-sleeve cotton shirt or welding jacket with long pants to protect skin. Remember to use welding curtains/screens and to have an extra welding helmet on hand for observers.

FIRE. Remove flammable items, such as lumber, rags, drop cloths, cigarette lighters, matches, and any other flammable items from the work area. Do not grind or weld in a sawdust-filled shop—sparks can ignite airborne dust and fumes or ignite flammable materials. Mount an ABC-rated fire extinguisher and first aid kit in your work area. Check your welding area for a full 30 minutes after welding to make sure no materials have been left to smolder or catch on fire.

EXPLOSION. *Never* cut, weld, or heat on a closed cylinder or container, because pressure could build and cause the container to rupture or explode. If this occurs, you may be responsible for the replacement cost of the cylinder if it was leased from a welding equipment supplier. Nonflammable and inert gases are stored in high-pressure cylinders with capacities of approximately 2,000 psi—**always** store these cylinders in a proper location with safety caps secured when not in use or during transport. If cylinders are in use with regulators attached, they must still be secured to a cart or structure. Never weld or strike an arc on a cylinder, because this could cause an explosion or rupture. Another explosion hazard is concrete. Always elevate material above a concrete floor when tacking or welding. The heat from the arc may cause the concrete to crack or explode.

ELECTRIC SHOCK. Electrical shock is an inherent risk when working with arc welding equipment. Always wear the proper PPE, making sure it is dry and free from holes. Never work in a wet environment, such as standing water, or while wearing wet gloves. Remember, electricity follows the path of least resistance, and water offers just that. Check equipment and welding leads daily to ensure insulation is not cracked or in disrepair; if such a defect is found, replace or repair the lead immediately to avoid serious electrical shock.

Other hazards include noise from grinding, cutting, sawing, and finishing metal; laceration from sharp metal edges; and asphyxiation from inert shielding gases such as carbon dioxide. **Always** read and follow the manufacturer's instructions before using any welding process.

MINIMUM LENS SHADE NUMBERS FOR WELDING

APPLICATION	SUGGESTED SHADE
Shielded Metal	
Arc Welding (SMAW)	
$^1/_{16}$ to $^5/_{32}$" electrodes	10
$^3/_{16}$ to $^1/_4$" electrodes	12
$^5/_{16}$ to $^3/_8$" electrodes	14
Gas Metal	
Arc Welding (GMAW)	
$^1/_{16}$ to $^5/_{32}$" non-ferrous	11
$^1/_{16}$ to $^5/_{32}$" ferrous	12
Gas Tungsten	
Arc Welding (GTAW)	10 to 14
Plasma Cutting	8
Oxyacetylene Welding	5
Oxyacetylene Cutting	5
Brazing	3 to 5

Welding helmets are typically available with filter lenses in either 2" × 4¼" or 4½" × 5¼" sizes. Helmets with auto-darkening lenses are also available. Clear, full-face protective shields are available for grinding or chipping and with a #5 filter for oxyacetylene operations.

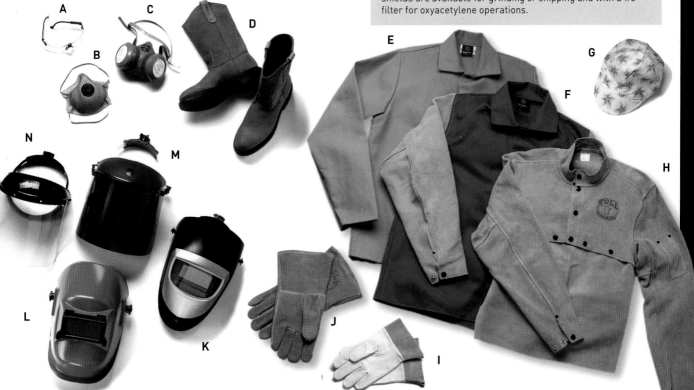

Welding safety equipment includes: **(A) safety glasses, (B) particle mask, (C) low-profile respirator, (D) leather slip-on boots, (E) fire-retardant jacket, (F) fire-retardant jacket with leather sleeves, (G) welding cap, (H) leather cape with apron, (I) leather gloves with gauntlets, (J) heavy-duty welding gloves, (K) welding helmet with auto-darkening lens, (L) welding helmet with flip-up lens, (M) full-face #5 filter, (N) full-face clear protective shield.**

METAL BASICS

The suitability of a metal for welding is dependent upon what elements are used to create the metal. It is important to identify the type of metal and what alloying elements have been added to increase its strength, toughness, impact resistance, corrosion resistance, or ductility. We typically divide metals into two categories: ferrous and nonferrous. Ferrous metals contain iron as their major element and include cast iron, forged steel, mild steel, and stainless steels. These metals are typically magnetic. Nonferrous metals include the "pure" elements, such as aluminum. Because they do not contain iron, they are nonmagnetic and have a lower melting temperature. It is essential that base metal and filler metal are matched to ensure that like metals are welded. Because this process does involve the melting and fusion of the base and filler into one new metal, melting temperatures and metal characteristics must be similar. We can join dissimilar metals by brazing, braze welding, or soldering, because these processes do not actually melt the base metal.

Ferrous metals contain iron with varying amounts of carbon and other alloying elements, such as chromium, molybdenum, manganese, and nickel. Each alloy imparts a unique characteristic to the low carbon steel. Mild steel (low carbon steel) is the most commonly used type and the easiest with which to work. Mild steel can be cut and welded with all of the processes covered in this book. It makes up most of the metal items you commonly use, make, or repair, including automobile bodies, bicycles, railings, furniture, cabinets, and shelving. Adding more carbon to the steel makes it harder but also more brittle and more difficult to cut or weld. These high-carbon steels are used to make cutting tools, such as drill bits, machining bits, and knife blades. Adding other alloying elements, such as chromium and nickel, to the low carbon steel produces stainless steel. Because the stainless steel does not oxidize (rust) easily, it is cut best with a plasma cutter. As always when welding low carbon steel or alloyed steel, the filler metal must be matched to the elements to ensure a high-quality weld is produced.

Aluminum is the most widely used nonferrous metal, because it is lightweight and corrosion resistant. Like steel, it is available in many alloys and is often heat-treated to increase strength. Aluminum is used for engine parts, boats, bicycles, furniture, kitchenware, and now automobile frames. Various characteristics make aluminum difficult to weld successfully—it does not change color when it melts, it conducts heat rapidly, and it immediately develops an oxide layer that melts at a higher temperature than the base metal itself, causing overheating and extreme distortion and metal destruction.

METAL	WELDING PROCESS	CUTTING PROCESS
Mild steel	All welding processes	Oxyfuel, plasma
Aluminum	Gas tungsten arc, gas metal arc	Plasma
Stainless steel	Gas tungsten arc, gas metal arc, shielded metal arc	Plasma
Chrome moly steel	Gas tungsten arc, oxyfuel	Plasma
Titanium	Gas tungsten arc	Plasma
Cast iron	Shielded metal arc, brazing	Plasma
Brass	Braze welding	Plasma

METAL MELTING POINTS

METAL	MELTING POINT (F)
Aluminum	1217°
Brass	1652 - 1724°
Bronze	1566 - 1832°
Chromium	3034°
Copper	1981°
Gold	1946°
Iron	2786°
Lead	621°
Mild steel	2462 - 2786°
Titanium	3263°
Tungsten	5432°
Zinc	786°

METAL SHAPES & SIZES

Mild steel and most other metals come in a variety of shapes, sizes, and thicknesses. Metal thickness may be given as a fraction of an inch, decimal, or gauge. Sheet metal is typically 3/16 inch or less in thickness and called out as a gauge number, and plate metal is 1/4 inch or thicker. Structural metal typically is identified by its length, width, and by its wall, leg, or web thickness. Some of the most common structure types and size call-outs are:

A. Rectangular tubing is used for structural framing, trailers, and furniture. Dimensions for rectangular tubing are specified by width × height × wall thickness × length.

B. Square tubing is used for structural framing, trailers, and furniture. Dimensions for square tubing are specified by width × height × wall thickness × length.

C. Rail cap is used for making handrails. Rail cap dimensions are the overall width and the widths of the channels on the underside.

D. Channel is often used for making handrails. Very large channel can be used in truck frames and structural items, such as bridges or industrial equipment. The legs, or flanges, make it stronger than flat bars. Dimensions for channel are specified by flange thickness × flange height × channel width (outside) × length.

E. Round tubing is not the same as pipe. Round tubing is used for structural items, while pipe is used to transport liquids or gases. Dimensions for round tubing are specified by outside diameter (OD) × wall thickness × length.

F. T-bar dimensions are given as width × height × thickness of flanges × height.

G. Angle or angle iron has many structural and decorative uses. Dimensions for angle are specified by flange thickness × flange width (leg) × flange height (leg) × length. Angle can be equal leg length or unequal leg length.

H. I. J. Square, round, or hexagonal bar (hex bar) dimensions are specified by width/outside diameter (flat to flat for hexagonal bar) × length.

Flat bar/strap (not pictured) is available in many sizes and is typically not wider than 12" for flat bar are specified by thickness × width × length.

Sheet metal (not pictured) is 3/16" or less in thickness, and is often referred to by its gauge number. Plate is 1/4" or greater in thickness, and is specified by its thickness in fractions of an inch.

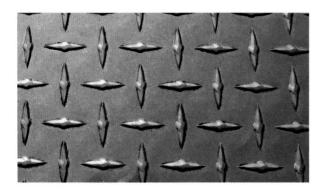

Diamond plate is used as a heavy-duty siding for toolboxes and carts because of its durability and strength. It is also used for decorative applications. It is sold as length × width × thickness.

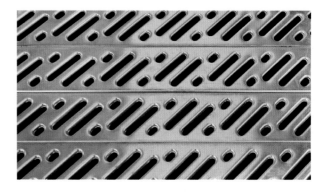

Perforated sheet metal is used for industrial shelving, fences, stair treads, grates for floor drains, and a host of decorative sidings and top treatments. There are numerous shapes and styles available. They are sometimes referred to as perforated metal screens. Perforated sheets are sold as length × width × thickness. Lighter-weight sheets are sold by the yard in a roll.

METAL IDENTIFICATION CHART

Test	Appearance	Fracture	Magnetic	Torch	Chip	Spark	Volume of stream
Manganese steel	Dull cast finish	Rough grained	Nonmagnetic	Turns bright red, melts quickly	Hard to chip	Bright white bursts, heavy pattern	Moderately large
Stainless steel	Bright, smooth surface lines	Bright appearance	Depends on exact composition	Turns bright red, melts quickly	Smooth chip, smooth bright color	Very few short full red sparks with few forks	Moderate
Low carbon steel (mild steel < 30% carbon)	Gray, fine	Gray, bright crystalline	Highly magnetic	Gives off sparks when melted, pool solidifies rapidly	Chips easily, smooth and long chip	Long white sparks, some forks near end of stream	Moderately large
Medium carbon steel (.30 to .45% carbon)	Gray finish	Light gray	Highly magnetic	Melts quickly, gives off some sparks	Chips easily, smooth and long chip	Long white sparks with secondary bursts along stream	Moderately large
High carbon steel (> 45% carbon)	Dark gray, smooth finish	Light gray-white, finer grained than low carbon steel	Highly magnetic	Melts quickly, molten metal is brighter than low carbon steel	Difficult to chip, brittle	Large volume of brilliant white sparks	Moderate
Wrought iron	Gray, fine surface lines	Fibrous structure, split in same direction of fibers	Highly magnetic	Melts quickly, slight tendency to spark	Chips easily, continuous chip	Straw-colored sparks near wheel, few white forks near stream end	Large
Cast iron	Rough, very dull gray	Brittle gray	Highly magnetic	Turns dull red, first puddle is very fluid, no sparks	Very small and brittle chips	Dull red sparks formed close to wheel	Small
High sulfur steel	Dark gray	Gray, very fine grain	Highly magnetic	Melts quickly, turns bright red before melting	Chips easily, smooth and long chips	Bright carrier lines with cigar-shaped swells	Large

PURCHASING METALS

Finding a metal supplier can be a challenging task. The materials that are readily available at home centers and hardware stores may not be the size and shapes needed for a project and can be very expensive; and if ordered online or through a catalog, product weight can create significant shipping charges. With some searching, though, most products can be found at reasonable prices. Once a metal supplier has been found, materials can typically be purchased by weight or per linear foot. The price for small pieces of mild steel at a home center or hardware store might be as much as $3 to $5 per pound, but the price at a larger steel supplier may be as little as $1 per pound, depending on the type of material ordered. Many steel suppliers have an odds and ends bin or rack where material is discounted even more than retail or wholesale prices. Stainless steel and aluminum are higher cost relative to mild steel. When purchasing at a home center or hardware store, material can be purchased in three-, four-, or even six-foot lengths that are much easier to handle. When purchasing from a steel supplier, the common lengths are 10, 12, and even 20 feet, so plan accordingly—as always convenience will cost more. Steel suppliers have most common types and

lengths in stock and can order other sizes. Some steel suppliers are distributors for decorative metal products, but many specialty items, such as wrought-iron railing materials, decorative accessories, and weldable hardware, are only available by catalog. A number of catalog supply companies sell to the public and have varied selections and reasonable prices (see Resources).

INCH EQUIVALENT FOR GAUGE THICKNESS

GAUGE	INCHES
24	0.020
22	0.026
20	0.032
18	0.043
16	0.054
14	0.069
12	0.098
11	0.113
10	0.128

Metal less than ⅛" thick is often referred to by gauge. For reference, the decimal equivalent of ⅛" is 0.125.

Wall plates, hooks, rings, balls, bushings, candle cups, drip plates, and stamped or cast items are available in a wide variety of shapes, sizes, and sheets.

METAL CLEANING & PREPARATION

A successful weld begins with a well-prepared welding joint. More attention paid while cleaning and preparing the welding joint ensures a higher quality weld with an acceptable appearance. When working with hot rolled steel (HRS), the mill scale—a thin layer of oxide formed when the material is processed while it is hot—needs to be removed. Clean all project parts to remove any oil, dirt, rust, and mill scale. This can be time consuming but will ensure a long-lasting project that will need little maintenance or repair. For any project that will be painted or powder coated, it is important to clean the entire project. If parts will be allowed to rust or become aged, thoroughly clean all areas to be welded.

The first step in cleaning is to remove all grease, oil, or dirt by wiping the part down with denatured alcohol, acetone, or a commercial degreaser. Alcohol works the best, as it has minimal odor and does not dissolve or damage plastics like acetone does. Both acetone and degreasers tend to leave a residue that may diminish the quality of the final weld or finish.

Once the grease, oil, and dirt have been removed from the surface, the mill scale must be removed. You can do this by wire brushing, grinding, sanding, or sand blasting the part. A bench-mounted grinder with a wire bush works well for cleaning small parts or the ends of smaller parts, but it cannot be used on larger surfaces. For larger surfaces, you will need an angle grinder outfitted with a grinding wheel, wire brush, or flap wheel. A hand-held, battery-powered drill with a wire brush or wire brush cup will also work. Remember, when working with any power tools, you must wear the appropriate PPE, including safety glasses, face shield, and long sleeves to protect your eyes and body from wire fragments that can be thrown from the brush. Also, never force or apply too much pressure to an angle grinder or drill, as the tool can kick back and cause severe injuries.

Once all parts have been cleaned and you have removed all of the mill scale, it is important to complete the project in a timely manner before the metal rusts.

Apply denatured alcohol with a clean rag to clean dirt and oil from project parts. Wear rubber gloves to protect your hands.

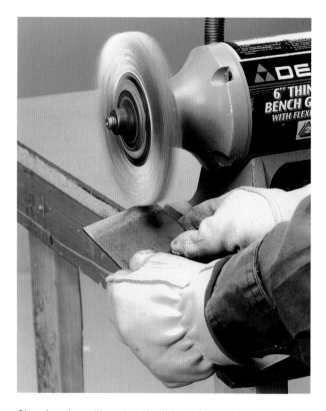

Cleaning the mill scale off mild steel is an important step. A bench-mounted power wire brush works well on small pieces. Wire brushing the entire project prior to finishing is critical for good paint adhesion.

SETTING UP THE WELDING SHOP

If you plan to weld on a regular basis, it makes sense to set up a welding shop or area in an existing shop. The primary concern, especially if you are going to share space with other interests, such as woodworking, will be to contain the hot sparks, slag, and other flammable elements that are naturally associated with welding. It is also imperative that all welding fumes be properly exhausted and that proper ventilation is provided to allow for a safe working environment. Although it is possible to weld outside, not all processes work well in an outdoor area. Wire feed welding (GMAW) and tig welding (GTAW) require a shielding gas, and wind can be a challenge, resulting in poor welds with unacceptable profiles or appearance. All of the arc welding processes must be performed in a dry environment to prevent electrical shock. In certain areas, a cold climate can negatively impact the welding process, as cold metals do not respond well to an electric arc. A heated garage or outbuilding is well suited for welding, while a basement is a poor choice due to the chance of fire and explosion, not to mention poor ventilation in or near a living space. Be

sure to check with your homeowner's insurance policy or landlord, as welding in an attached garage may void your policy or lease.

Remember that very small sparks and pieces of slag can scatter or be projected from a grinder or welder up to 40 feet from the source. If they land on a flammable or combustible material they may smolder, and under the right conditions they can ignite. Always prepare for cutting, welding, and grinding operations with the proper protective welding curtains and remove all combustible material from the immediate area. Perform a safety walk before and after welding to make sure the area is clear and that no material is smoldering. Never weld on wooden structures or wooden structures covered with metal, as the heat may transfer from the metal through the wood, which can smolder for a long period of time before igniting.

Shop space with a concrete floor and cement block walls is ideal for welding. Good ventilation is also important.

SHOP TOOLS

Selecting tools for the welding shop can be a daunting task. The wish to buy cheap while expecting quality sometimes frustrates new welders. Invest as much as you can comfortably spend, and buy industrial grade equipment from a welding supplier. If this means spreading your purchases out over a long period of time, that is okay—you will be more successful and less frustrated in the end. Beware of buying used welding equipment online. Some older machines may work great, and the price is usually right. But as newer models come out, finding consumable items such as electrode holders, GMAW liners, and repair parts can be challenging. Do your homework and research the availability of parts, consumables, and accessories for the machine before buying. Remember this is one of the largest investments you will make in your welding career.

Now that you have a welder you can purchase power tools. An angle grinder, portable bandsaw, and a chop saw are very useful for cutting and preparing parts for welding. A drill press and bandsaw can also be useful.

As your welding skills and fabrication knowledge improve, you can add specific metalworking equipment, including metal brakes and benders, tubing benders, and scroll benders for ornamental work. These pieces of equipment can range in price depending upon the name brand, size, and whether they are manually or hydraulically operated.

Standard tools for a welding shop include:
(A) hacksaw, (B) combination square,
(C) ball peen hammer, (D) framing square,
(E) various sizes of C-clamps, (F) center punches, (G) alternative C-clamp, (H) measuring tape,
(I) magnetic clamp, (J) cold chisel, (K) circular saw, (L) portable band saw, (M) air compressor, (N) angle grinder,
(O) angle pointer/calculator, (P) level, (Q) metal file, (R) oxyacetylene and oxygen (including regulators, hoses, torch with tip), (S) plasma cutter, (T) TIG stick welder, (U) multiple-purpose cut-off saw, (V) MIG welder.

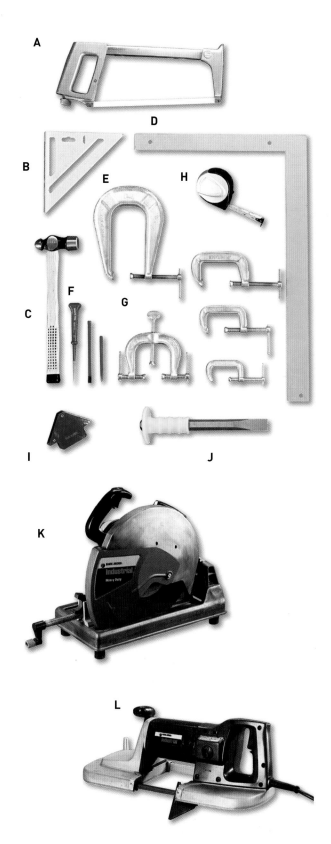

M

N

O

P

Q

R

S

T

U

V

METAL REPAIR

Once you begin welding, you will encounter numerous opportunities to repair items. When your friends and neighbors discover you can do repairs, even more challenges will come your way.

Performing weld repairs can be very tricky, even dangerous, if you are not sure of the type of material and what the part was used for. When in doubt, bow out, especially on a vessel or drum where you are unsure of its previous contents. It is important to assess your welding skills, the difficulty of the repair, and the intended use of the repaired item. Any structural or vehicle repairs, such as stairways, ladders, trailers, or chassis, need to meet the same safety standards as they did in the original condition.

The first step when considering a repair is determining why the item broke or failed. If the failure was due to a poorly executed weld, the repair might be as simple as grinding out the welded area, tacking the area back together, and performing a satisfactory weld. But if a piece has broken due to metal fatigue, simply welding over the cracked area may only cause more cracks in different areas. Investigate and ask questions, and reach out to mentors with suggestions of how to properly repair the failed item.

The second step in a repair is determining the base metal. A magnet will be attracted to metals with a fairly high concentration of iron, but stainless steel (which is sometimes nonmagnetic), mild steel, and cast iron each require very different welding techniques. Aluminum is nonmagnetic and is discernable from stainless steel by its light weight—but which alloy is included with the aluminum? Some aluminum alloys are not weldable. Unfortunately, many manufactured metal items are alloyed and may have been heat-treated. Without access to the manufacturer's specifications, it is sometimes impossible to determine the composition of the base metal. It is important to understand the effects of welding on these materials before attempting a repair.

The third step after determining the feasibility of a repair is to properly clean and prepare the area to be welded, removing all dirt, oil, and paint or finishes. If one of the arc welding processes is selected, you must also clean an area down to bare metal so the work clamp may be properly secured. If the break is at a welded joint, you must also remove areas of the old weld bead to ensure complete penetration into the base metal.

All parts need to be carefully prepared before attempting a repair. Paint needs to be ground or sanded off and grease and oil need to be cleaned away. This cast iron part is being beveled to allow greater weld penetration.

Mild steel is the easiest material to repair. Simply prepare the metal as we discussed in the metal cleaning and preparation section and the weld may then be completed with any of the arc welding processes.

Cast iron and cast aluminum both need to be preheated before welding to help prevent cracking due to temperature fluctuations. If the piece is small enough, it can be placed in an oven at 400 to 500 degrees Fahrenheit. Otherwise, use an oxyacetylene or oxy-propane rosebud or heating tip. Temperature crayons that melt at specific temperatures are available for marking metals for preheating. Post-weld heating and penning may also be required to dissipate stress while the part returns to ambient temperature at a controlled rate—this will further reduce the chance of cracking.

Some aluminum is not weldable, but if an aluminum part has been welded before, it is typically safe to perform a repair weld. If the composition of the base metal can be determined, match the filler metal makeup to the base metal. Some filler metals are multipurpose and can be used on more than one alloy. Consult the manufacturer's recommendations or ask a professional at a welding supplier.

Stainless steel also comes in numerous alloys, and it is important to match the filler metal with the base metal. Do not clean stainless steel with a wire brush, as the mild steel wires may contaminate the base metal. Instead, use a flap wheel, emery cloth, or an abrasive pad.

Remove the paint and finish from the weld area. If you are arc welding, also remove the paint from an area close to the weld for attaching the work clamp.

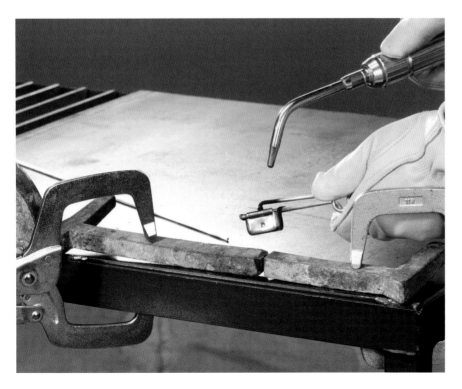

Cast iron can be braze welded as shown here, or shielded metal arc welded with cast iron electrodes. Either way, the metal needs to be preheated to 450°F before being welded.

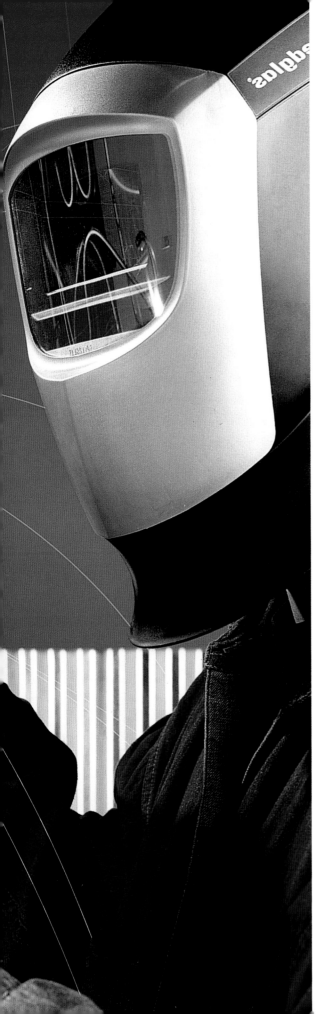

TECHNIQUES

In this chapter we will discuss the basic cutting and welding processes. These techniques are the foundation of your skill set as a welder—skills you will use again and again, regardless of the type of welding you pursue. This chapter also provides basic directions and step-by-step photos that illustrate major welding and cutting processes. You will find quick reference charts that describe electrode selection and filler wire choices, metal types and weldability, and joint design and weld types.

MECHANICAL CUTTING

For some projects, cutting the metal is the most difficult part. A variety of tools exist for cutting metal, but as the thickness and size increases, the choices become limited. A simple hacksaw can be used for cutting tubing, solid rod, or solid bar stock, but larger material with thicker walls will have you looking for power cutting equipment. Cutting thin-gauge sheet metal can be accomplished with hand sheetmetal snips, but plate steel will require a mechanical or hydraulic-powered shear or a cutting torch. You can spend as little as $100 or as much as several thousand dollars on metal cutting equipment, so carefully consider the options.

When choosing to cut with a blade, look for bimetal saw blades, whether in a hacksaw or bandsaw. To prolong the life of your blades, do not apply excess force or downward pressure, but rather allow the blade to cut while applying slow, steady pressure. Forcing the blades into the material simply wears the blade out prematurely.

A horizontal metal-cutting bandsaw is a bench-mounted saw with clamps to hold work pieces, and an automatic shut-off feature turns off the saw when the cut is completed. The most common cutting capacity is 4×6 inches, which can cut through a rectangular piece of that dimension or a 4½-inch round. A portable metal-cutting bandsaw is slightly less expensive than the bench version and obviously more portable. The most common size cuts 4-inch stock. It is more difficult to create accurate cuts with a portable bandsaw, but the portability makes the machine more versatile in a do-it-yourself shop. A very distinct advantage of a bandsaw is the very small kerf, usually less than ⅛ inch.

Metal-cutting chop saws and angle grinders use abrasive wheels for cutting. Most of these produce a great deal of heat, dust, and sparks, and the cut pieces typically have burred edges that require more prep work before welding. Metal-cutting chop saws with carbide-tipped blades produce cleaner cuts with less heat. They also cut faster than most abrasive chop saws. Both types of saws produce a larger kerf than the bandsaw—5/32 inch or more. These saws operate like a circular saw or chop saw used to cut wood, but can cut ¼-inch steel and all types of tubing and solid bar.

Manual or hydraulic metal shears in small sizes are suitable for home shops and are relatively inexpensive.

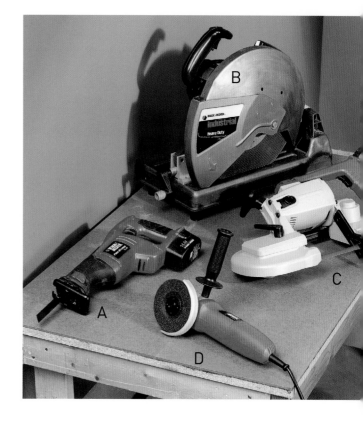

Power tools for a welding shop include:
A   reciprocating saw
B   cut-off saw
C   portable band saw
D   angle grinder

They are typically limited to use with thin-gauge sheet metal, usually less than 16 gauge. Hydraulic shears capable of cutting metal ³/₁₆ of an inch or thicker are more expensive.

Manual or hydraulic metal shears and punches in small sizes suitable for home shops are relatively inexpensive. They are limited to thin-gauge materials, usually less than 16 gauge. And the heavier the material capacity is, the more expensive the machine.

Manual snips are sufficient for cutting curves in sheet metal thinner than 18 gauge. Power nibblers make quick work of curves but are commonly available for 18-gauge or thicker.

PLASMA ARC CUTTING (PAC)

Plasma cutters work by sending an electric arc through a gas that is passing through a constricted opening. The gas can be shop air, nitrogen, argon, oxygen, etc. This elevates the temperature of the gas to the point that it enters a fourth state of matter, known as *plasma*. Because the metal being cut is part of the circuit, the electrical conductivity of the plasma causes the arc to transfer to the work. The restricted opening (nozzle), which the gas passes through, causes it to squeeze by at a high speed, like air passing through a venturi in a carburetor. This high-speed gas cuts through the molten metal. The gas is also directed around the perimeter of the cutting area to shield the cut. Unlike the oxyacetylene process, which is only used to cut oxidizable parts, the plasma-cutting process can be used to cut any conductive metal—aluminum, brass, cast iron, copper, steel, stainless steel, and titanium.

Though the plasma arc temperature is 40,000 degrees Fahrenheit, it is so constricted and the cutting speed is so fast that the thermal distortion to the metal being cut is

very minimal. Plasma cuts are clean and often weldable with little or no additional cleanup, if the base metal was initially clean. Performed properly, the kerf it produces is a straight-sided cut with very little to no slag present.

Plasma cutting safety. Plasma cutting produces sparks, smoke, fumes, and ultraviolet light. It is important to wear the proper eye and face protection; shade-8 lens is recommended. Proper ventilation is also required, as cutting materials that contain chromium, nickel, or that are galvanized produce fumes that can be toxic. Never cut through paint or oil/dirty material, because as they burn off during the cutting process they can produce hazardous smoke.

A plasma cutter consists of a power source with a compressed air or inert gas connection, work cable with clamp, supply cable, and torch.

A dehumidifying filter is very important for the compressed air source for plasma cutting.

Because of the high level of open-circuit voltage, plasma cutting has an added electrical shock hazard. Make sure all surfaces around the cutting area, as well as your clothing and gloves, are dry. Turn off the power to the machine before changing any torch parts. **Never** cut containers, tanks, or cylinders that may have held flammable materials. Even a small amount of flammable residue can cause fumes to accumulate inside an enclosed vessel, causing an explosion when cut.

Equipment. Plasma-cutting machines can be expensive, costing several thousand dollars, but as technology advances are made and demand increases, companies are introducing smaller, less expensive small shop equipment. If you regularly need to cut steel, stainless steel, and aluminum, this machine can be worth the cost. Small, 115-volt plasma cutters are available that will cut up to ⅜" material. Larger 220-volt machines can cut thicker materials and have a higher duty cycle. Aluminum and stainless steel require higher amperages for cutting, so check to make sure the machine can cut the materials with which you typically work.

Most plasma cutters require a compressed air source. Any shop air compressor that can deliver a constant 65 to 80 psi will be sufficient. It is critical that a filter and or water separator unit be installed to keep the air dry and oil free. Follow the manufacturer's directions for installation and placement of the unit. Some plasma cutters have an internal air compressor—these units are more expensive and typically require more maintenance.

Most plasma torches have a manual switch on the torch body (handle). The cutting tip, electrode, and nozzle are consumable parts, so check these parts frequently for wear and replace them when necessary.

The plasma cutting torch consists of (left to right) a shield cup, cutting tip, starter cartridge, electrode, and torch handle.

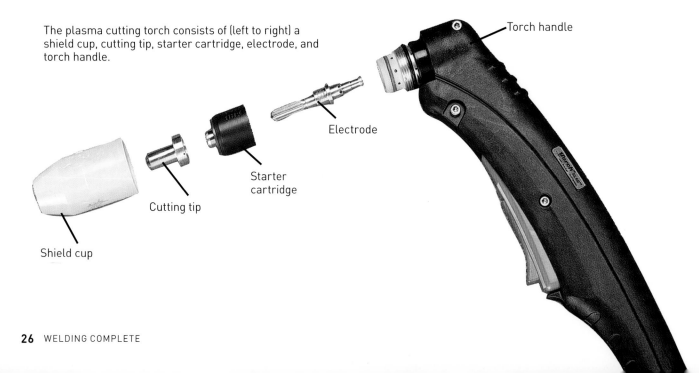

Torch handle

Electrode

Starter cartridge

Cutting tip

Shield cup

HOW TO PLASMA CUT

1 Check the manufacturer's recommended settings for the material to be cut. Arrange the material on a cutting table and determine the most comfortable cutting position. Attach the work clamp to the workpiece or welding table. Dry run through the cutting motion to practice speed and cutting position.

2 Check manufacturer's directions for arc starting procedure. Most current machines are designed to drag the nozzle directly on the material's surface. Activate the arc and hold the torch perpendicular to the surface of the material. Move steadily and smoothly along the entire length of the cut. Do not stop the arc until you have passed through the end of the cut.

PRECUTTING CHECKLIST

- Compressed air fittings are tightly attached.
- Cables are in good condition.
- Ventilation is sufficient to remove fumes from cutting area.
- Work area is dry.
- Flammable materials are removed from area.
- Sparks and slag from cutting will be contained.

3 A good plasma cut has squared edges, small vertical ripples, and little or no slag.

TROUBLESHOOTING & TECHNIQUES

A high-quality circle-cutting attachment for a plasma cutter is expensive, but worth the investment if you will be making repeated circle cuts. Inset: This unit comes with a magnetic pivot holder for steel and a suction cup pivot holder for all other metals.

PLASMA TROUBLESHOOTING

PROBLEM	SOLUTIONS
Arc does not start.	• Connect work clamp to workpiece. • Turn on power source. • Tighten cable connections. • Make sure electrode is in working condition.
Cut does not penetrate through material.	• Cut more slowly. • Make sure torch is held perpendicular to material. • Make sure machine is capable of cutting that thickness. • Increase amperage setting. • Verify correct air flow and pressure.
Excessive slag or dross is generated.	• Increase cutting speed. • Decrease torch standoff distance. • Replace worn torch parts. • Adjust amperage setting.
Torch parts are consumed quickly.	• Install or replace air filter to prevent oil or water from reaching torch. • Cut metal within the capability of the machine. • Make sure gas pressure is correct.

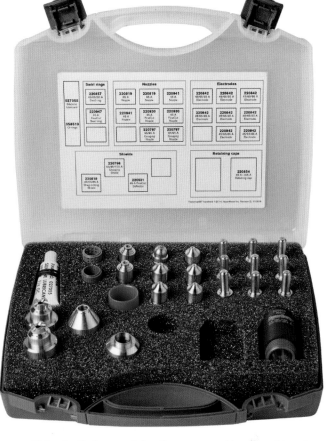

The adjustable pressure regulator with gauge is located on the front of the Lincoln Pro-Cut 55 plasma-cutting machine. This compact machine generates 60 amps of cutting power capable of cutting conductive materials up to ¾" thick. It has a built-in air filter.

This plastic box contains the consumables for a Lincoln plasma cutter. The consumables are made of copper, which resists sticking to almost all other metals. The parts should be replaced when they become too worn out to work properly.

TIP

Making a good-quality plasma cut is dependent on a consistent travel speed. Consult the owner's manual for cutting speeds for different material thicknesses and amperage settings. With the machine off, practice moving across the planned cut at the correct speed.

MORE PLASMA CUTTING TECHNIQUES

This operator can make precise cuts with a plasma cutter without adding the heat input of an oxyacetylene torch. This will prevent heat distortion in surrounding regions of the work area.

Plasma cutters can also pierce small pieces of thick metal, but for a cleaner cut first drill a small, ⅛" hole to start the cut.

BENDING & CUTTING DIAMOND PLATE

Although diamond plate can be plasma cut, there are times when you'll want to bend the sides instead. Cut a bird's mouth with a plasma cutter (see page 97) and then use a simple break such as the one shown here to bend the sheet (see page 101). Bend a little at a time until you get what you want. If you score the backside beforehand it makes the bend easier with less stress, but don't score it deep or it will crack.

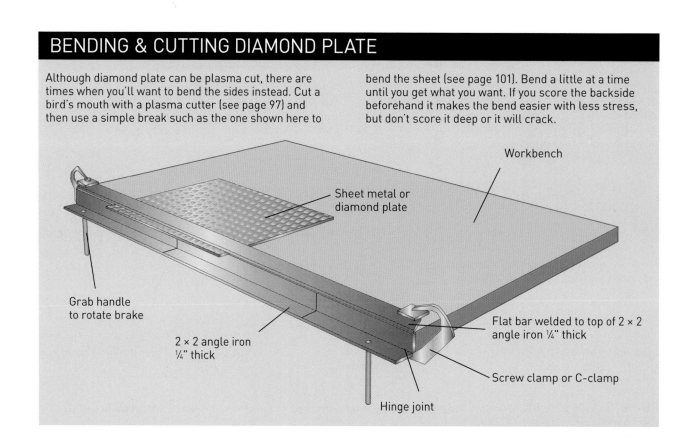

Workbench

Sheet metal or diamond plate

Grab handle to rotate brake

2 × 2 angle iron ¼" thick

Flat bar welded to top of 2 × 2 angle iron ¼" thick

Screw clamp or C-clamp

Hinge joint

Cutting pipes is a cinch with an electric plasma cutter.

Making detail cuts is easy with a plasma cutter. Here the welder is cutting scrap metal for recycling.

Here, custom motorcycle builder Scott Webster (of Leroy Thompson choppers) uses the Pro-Cut 55 to cut a piece of steel over a barrel that catches the debris.

The Lincoln Pro-Cut 25 plasma cutting machine is good for fish-mouth fitting tubing, including tailpipe tubing. After the cut is made, a die grinder removes slag from the cut line.

A Miller plasma cutter is being operated by a gasoline generator and an air compressor to make a repair on this hay baler. When the broken part is removed, a replacement will be welded on by a wire-feed machine.

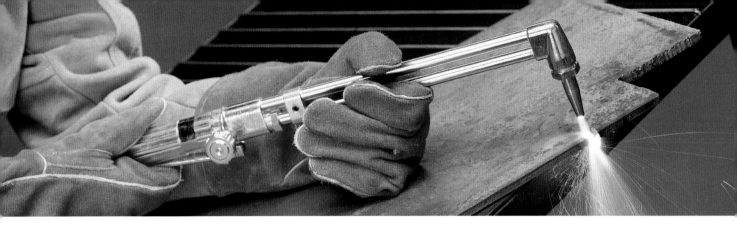

OXYACETYLENE CUTTING (OAC)

Oxyacetylene cutting uses oxygen and acetylene to preheat metal to a temperature of approximately 1,600 degrees Fahrenheit, and then uses a stream of pure oxygen to oxidize or burn away the molten metal. Because the cutting is achieved by oxidation of the metal, oxyacetylene cutting only works on metals that are readily oxidized at this temperature. Metals that can be cut with this process include mild steel and low alloy steels, with thicknesses ranging from ¼" to 12". Oxyacetylene cutting is relatively inexpensive when compared to plasma cutting, and it is very portable because no external power source is needed. However, OAC is limited to steel. Used on steel, the molten slag and sparks present a fire and burn hazard, and the heat input to the base metal may cause distortion or changes in the properties of the steel.

Safety. The risk of starting a fire or being burned is high with oxyacetylene cutting. It is important to prevent molten slag and sparks from coming in contact with skin and flammable materials. Make sure to wear heavy leather gloves with gauntlets, a leather jacket or leather-sleeved jacket, and leather boots. The work area should be cleared of flammable materials. If cutting an item in place, make sure the surrounding areas are protected from heat, sparks, and dropping slag. Never cut sealed tanks, cylinders, or items that have contained flammable material. Never cut near fuel tanks or lines.

Equipment. The basic equipment for oxyacetylene cutting is the same as for oxyacetylene welding, with the addition of a cutting torch or cutting-torch attachment. The setup consists of a high-pressure (2,200 psi) oxygen cylinder, a low-pressure (225 psi) acetylene cylinder, regulators designated for oxygen and acetylene, hoses, and a torch with a cutting attachment or dedicated cutting torch. Cutting attachments and dedicated cutting torches have levers to activate the oxygen flow used to remove the molten metal.

SAFETY

+ Full-face #5 filter
+ Leather, wool, or cotton long pants, leather jacket, and hat
+ Heavy-duty welding gloves
+ Leather boots or shoes
+ Ventilation

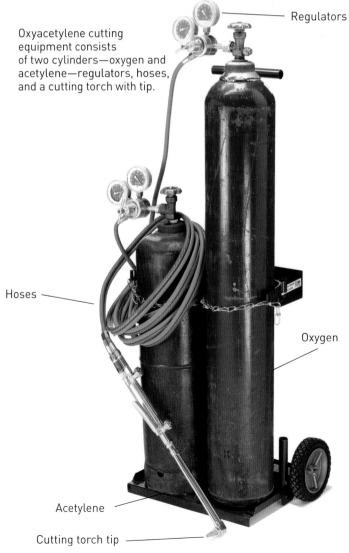

Oxyacetylene cutting equipment consists of two cylinders—oxygen and acetylene—regulators, hoses, and a cutting torch with tip.

Regulators

Hoses

Oxygen

Acetylene

Cutting torch tip

Cutting torch & tips. Most oxyacetylene welding/cutting sets come with a cutting attachment with an assortment of cutting tips. The cutting attachment attaches to the handle just as the welding tips attach. This is called a combination torch. A dedicated cutting torch is a one-piece unit that has its own mixing chamber. It is greater in length and allows for more distance between the heat zone and the operator, and it can handle higher oxygen flow rates. A dedicated cutting torch is usually fairly expensive and not necessary unless you are cutting very thick metals, which typically require higher oxygen flow.

Both the dedicated torch and cutting attachment take a variety of cutting tip sizes. The cutting tip has a number of preheat holes surrounding the pure oxygen orifice. It is important to match the cutting tip to the thickness of the metal being cut. Avoid the one-size-fits-all approach to cutting tips, as this can lead to poor cuts with heavy slag, which requires more post-cut cleanup. Thinner metals require a tip with fewer preheat holes and a smaller oxygen orifice, while thicker metals require more preheat holes and a larger oxygen orifice. Cutting tips are made of copper and can be damaged easily; clean them regularly with a tip cleaner of appropriate size.

Fuel gas options. Because the metal being cut does not need to be brought to the melting point, various gases other than acetylene can be used. Propane, natural gas, propylene, and methylacetylene-propadiene can be used

A striker (A), tank wrench (B), pliers (C), and tip cleaner (D) are needed for oxyacetylene cutting.

as the fuel source. Each of these gases requires specific regulators and specific cutting tips designed for each specific gas. **Only use tips designed for each gas**—never utilize a tip designed for propane to burn acetylene, because catastrophic results will occur. Check the manufacturer's recommended usage before selecting a tip for a specific gas.

A dedicated cutting torch (top) is a one-piece unit with one oxygen and one fuel-gas valve. A cutting tip is attached to a standard torch body (bottom) and includes an additional oxygen valve. A range of tip sizes is available for both torch styles.

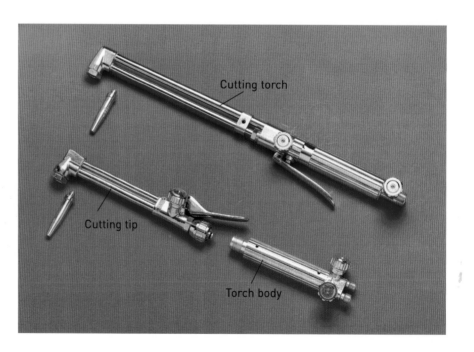

Cutting torch

Cutting tip

Torch body

HOW TO CUT WITH OXYACETYLENE

1 See pages 44 to 47 to learn how to set up, pressurize, and light an oxyacetylene torch. Set up the material to be cut on a cutting table. Use soapstone to mark the cutting line, and practice your cutting position and bracing. Hold the torch in your left hand with your thumb and forefinger on the acetylene valve. Hold the striker in your right hand, 3 to 6" from the torch tip. Turn the acetylene valve ¼ to ½ turn and strike sparks.

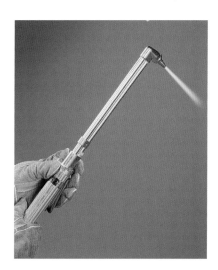

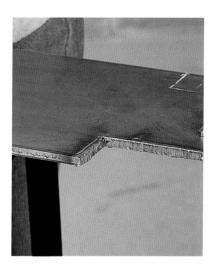

2 Adjust the acetylene so the flame is not smoking and is not separated from the tip (see page 47). Add oxygen to create a neutral flame. With a dedicated cutting torch, depress the oxygen lever to make sure neutral flame is maintained; if necessary, adjust with the oxygen lever depressed. With a cutting tip on a torch base, depress the oxygen lever and use the cutting tip adjustment valve to adjust the oxygen.

3 Release the oxygen lever, flip down your face shield, and direct the flames at the edge of the cut. Once the metal begins to glow red and appears shiny but not yet molten, depress the oxygen lever. If it fails to pierce a hole immediately, release and continue to heat until depressing the lever pierces a hole. Move slowly and steadily along the cutting line, holding the torch at a 90° angle.

4 The finished cut should have very small vertical ridges (draglines), very little slag on the bottom, and the top edges should not be rounded over.

OXYACETYLENE CUTTING TIPS

PIERCING HOLES AND MAKING CUTOUTS

Piercing a hole: Preheat the material to a dull red. Pull back the torch from the surface and angle slightly. Squeeze the oxygen lever. As soon as the material has been pierced, return the torch tip to the perpendicular position and move back to just above the surface. Complete the hole.

Making a cutout: Begin in the center of the cutout and move to the cutout line after piercing a hole. A spiral approach to the edge works best. To give maximum protection to the surrounding material when piercing a hole, drill a ³⁄₁₆" hole to start the piercing action.

WARNING

- Never cut into a sealed container.
- Never cut into a tank or container that contains or may have contained flammable materials.
- Never cut near a gasoline tank or fuel lines.

Oxyacetylene cutting safety.

Oxyacetylene cutting produces molten slag, sparks, and hot metal scraps. Move items that are to be cut away from burnable material. If the item cannot be moved, protect surrounding areas with sheet metal and fire retardant welding blankets (not tarps). Make sure sparks and slag will not fall into cracks, holes, or ventilation grates in the floor. Wet down any wooden material around the cutting area. Have water, sand, or a fire extinguisher on hand, and monitor the area for one half hour after cutting is completed. Plan where cut metal will fall so it does not hit your arms, legs, gas hoses, or cylinders.

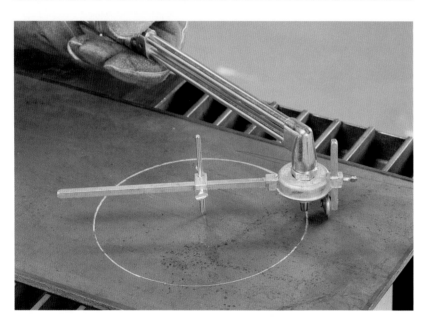

Circle cutting guides are available for making accurate oxyacetylene circle cuts. If you need to make several circle cutouts, this can be a handy tool.

CUTTING DEFECTS

A good oxyacetylene cut has square edges, small vertical draglines, and little slag. A dirty cutting tip is the cause of many poor-quality cuts. Clean the orifices with a properly-sized tip cleaner and use very fine sandpaper on a flat surface to polish the flat surface of the tip. Here are a few possibilities for causes of cutting defects:

- Kerf is wider at the bottom than at the top (bell-mouthed).
- Oxygen pressure is too high.
- Top edge melted over.
- Cutting speed is too slow or preheat flames are too long.
- Irregular cut edge.
- Speed is too fast or too slow, oxygen pressure is too high, or cutting tip is too small or too large.

Use a piece of angle iron as a brace to support the torch or to hold the pipe in place for cutting.

SHAPING

Bending and shaping mild steel can be done in a number of ways. Brakes for creating or bending angles in sheet metal, and ring rollers and scroll benders to create circles and scrolls, are readily available for the home shop. Unfortunately, the higher the capacity of these bending tools, the higher the price. But you can make simple bending jigs from any round, rigid, strong forms of metal, such as pipe, salvaged flywheels, or pulleys. Using these jigs, it is relatively easy to bend a round rod up to ¼ inch in diameter and a flat bar up to ⅛ inch thick into complex shapes. A bench vise is handy for making acute bends. Tubing, rail caps, and channels of minimal thickness can be bent with a heavy-duty conduit bender. With practice and patience, you can create well-formed circles by making incremental bends with a conduit bender.

Use a flat bar and locking pliers to bend sheet metal.

Bending jigs can be made from many rigid, circular items. An engine flywheel (A), toilet flange (B), and various pipe sizes (C) are shown here.

Use locking pliers to hold the metal to the jig. Wrap the metal around the jig in a smooth motion.

Use a scroll bender to bend decorative scrolls.

FINISHING

A nice finish adds to the beauty of your welded project. Unless you intend to let your project fully rust, even bare metal needs some type of coating. Applying paint coating is best done by spraying, not brushing.

If you look closely at most welded furniture or garden accessories, you will see that they typically have rough welds and spatter. With your own projects, especially indoor projects, take the time to grind and smooth welds and remove any spatter to increase the aesthetic value of your projects. This is especially important if you plan to apply a glossy finish, because a smooth, shiny paint job will highlight even the smallest imperfection.

A primer coat is recommended underneath all but bare metal coatings, such as polyurethane clear finish. If areas that have already rusted are hard to clean, use a primer designed for rusted metal or a rust converter. Carefully read the product label, as some converters require oil-based top coats and other specific finishes.

A wide variety of spray-on finishes is available. With a single application, you can create beautiful, detailed faux finishes, such as a hammered texture or granite. Crackle and antiquing finishes are also available as single spray-on finish paints and brush-on paints. Of course, you can create your own antique look as well (see photo, top right).

If you have more time and a higher budget, consider a verdigris finish. This finish requires more steps and time due to paint layers, but it creates a unique look that is

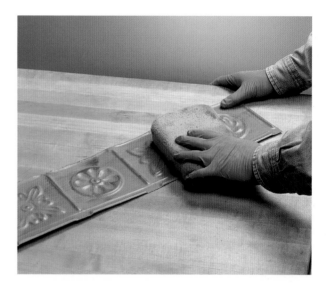

One method for creating an antique finish is to apply two wet top coats of different colors to a dry base coat of a third color. Before the top coats dry, wipe the top coats off in selected areas with a wet sponge.

There are many possibilities for finishing metal. Pictured here, from left to right: brush-on crackle finish, spray-painted stone, spray-painted hammered metal, and a swirl pattern created with an angle grinder.

An angle grinder works well for beveling edges on thick material. Grinding down welds gives projects a finished look.

very impressive, yet natural looking. The great advantage of simulating a patina is that you can control the hues and coloration, and then seal it all so that it remains perfect for years and years. Powder coating and brass plating are other, more expensive, finishing options.

For exterior applications, look to railing and gate suppliers; they carry a variety of primers, paints, dyes, and patinas to give your project a one-of-a-kind look and the ability to withstand constant exposure to natural elements. If the product will be near open flames or heat, make sure to buy paint specifically for such an environment.

Never to be underestimated, bare metal creates intriguing looks just by applying sanding, grinding, and heating techniques. One example is to use an angle grinder with a flap disc, grinding in a circular motion to create a unique pattern and texture. It can be time-consuming work that requires a little elbow grease and muscle, but the finish is unmatched.

Creating an antique finish can be accomplished in a number of ways. One is to paint the surface with three or four layers of different paint colors, allowing each layer to dry thoroughly between coats. When the final coat is dry, sand off corners and high spots to reveal the different paint layers. Another method is to apply several layers of paint colors without allowing the layers to dry between applications. Then use a wet sponge to wipe off layers and corners and high spots. To create a crackly "alligator" appearance, purchase crackle base and crackle top coat, either as a spray paint or brush-on paint. Apply two coats as directed to create the crackle effect.

Use an angle grinder in a circular motion to achieve a shiny textured effect. On thin material, as shown here, use small circles with light pressure.

Clean and then apply primer to prevent rust. Standard white is suitable if you intend to paint the finished project.

After the primer is completely dry, add the final top coat paint. This top coat should also be formulated for metal and/or exterior applications.

OXYACETYLENE WELDING (OAW)

Oxyacetylene welding is a process that uses heat from a gas flame to melt base materials and create fusion at the welding joint. This type of welding is autogenous, joining metal without the addition of filler metal. The more common approach is to use filler metal. The flame is created by the combustion of oxygen and acetylene in a precise mixture or volume. Other fuel gases can be used, including propane, hydrogen, natural gas, and methyl acetylene-propadiene (MPS, formerly MAPP gas), but the most common for welding is acetylene. It is the chosen fuel because oxygen and acetylene burn in a neutral flame at a temperature between 5,600 and 6,300 degrees Fahrenheit, the hottest of any gas flame and capable of melting most metals. Other oxygen-fuel gas mixtures are not capable of producing the necessary BTUs for welding, but they can be used for soldering or brazing.

The oxyacetylene process is versatile, as it can be used for both welding and cutting metals and also for heating, soldering, and brazing. It can be much less expensive than arc welding and is very portable, because it needs no electrical power source. It can also be used to weld a variety of thicknesses, though sections over ¼ inch are difficult. Unfortunately, the techniques of oxyacetylene welding can be very hard to master, and there are serious safety concerns with the extreme flammability of acetylene. Oxyacetylene welding equipment consists of an oxygen cylinder, acetylene cylinder, pressure regulators, hoses, torch, and welding tip.

An oxyacetylene welding rig consists of two cylinders—oxygen and acetylene–regulators, hoses, and a torch with tip.

THE EQUIPMENT

Cylinders. Oxygen is stored in a high-pressure cylinder (sometimes called a bottle or tank) at 2,200 pounds per square inch (psi) when full. The cylinder valve should always be covered with the provided safety cap when stored or during transport. Never lift the cylinder by the safety cap. Always store and transport the cylinders in the upright position.

Acetylene is an extremely unstable gas. It should **never** be pressurized above 15 psi in its free state. To safely pressurize the acetylene to 225 psi in a cylinder, the acetylene cylinder is filled with a porous material and acetone. When pressurized, the acetylene is absorbed by the acetone, which provides a stable medium for the gas. When the pressure is released (the valve is opened), the acetylene bubbles out of solution. Acetylene cylinders have fusible safety plugs that will melt in case of a fire and allow the gas to slowly release, not explode. Acetylene cylinders should always be stored in an upright position. Using an acetylene cylinder at an angle will allow acetone through the regulator and hose, which can damage them.

Cylinders should always be kept upright and secured with a chain or strap to prevent them from failing over. Never use a cylinder as a roller for moving heavy objects or for any purpose other than its intended use.

Cylinder valves. Each type of cylinder has a valve that controls the flow of gas from the cylinder, operated by a hand wheel or valve wrench. High-pressure cylinders, such as oxygen and argon, should always be opened fully to backseat the valve—the valve is leak-proof when completely closed or completely opened, but not otherwise. Cylinder valves are designed for specific gases. Each cylinder valve is threaded for a specific regulator—this is a safety feature to ensure that the proper regulator is used for each gas.

The acetylene valve should never be opened more than ¾ to 1½ turns. Any less may lead to insufficient gas flow and a backflash; more makes it difficult to turn the cylinder valve off quickly in case of an emergency. If the acetylene valve is opened all the way or opened too quickly, the acetone may escape from the cylinder and may destroy the hoses and regulator diaphragms.

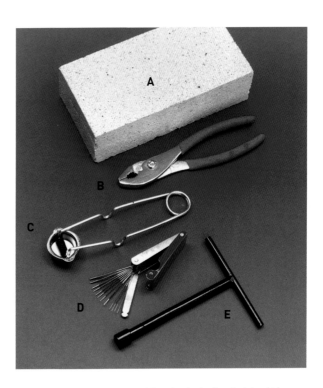

Tools for oxyacetylene welding include fire bricks (A), pliers (B), striker (C), tip cleaner (D), and acetylene tank wrench (E).

FIRE BRICKS

Fire bricks are specially made bricks that have little water content so they will not explode when heated to high temperatures. Fire bricks do not pull the heat from your weld or braze, and you cannot weld your material to them.

Pressure regulators & gauges. Pressure regulators reduce the pressure of the gas leaving the cylinder. A single-stage regulator reduces the cylinder pressure to a working pressure in one step, while a two-stage regulator reduces the pressure in two steps. A two-stage regulator can provide a more precise gas flow, and is considerably more expensive than a single-stage regulator, though a single-stage is sufficient for most welding conditions. To adjust, turn the regulator screw clockwise to increase pressure and counterclockwise to decrease. The regulator is completely off when the adjustment screw is loose or "backed out." Always back out the adjusting screws as part of the post-welding routine.

Hoses. The flexible rubber hoses that move the gas from the regulator to the torch are designed to be leak-proof and to withstand high pressure. The oxygen hose is green, and the acetylene or fuel hose is red. Hoses are available as a single hose or dual hose (twin), in which the fuel and oxygen hoses are paired together. Twin hoses are the most common and more convenient. Always protect hoses from damage by moving them off of the floor when not in use, keeping them behind the welding area so they are not exposed to sparks, and keeping them out of traffic lanes or areas so they are not stepped on or run over by vehicles. Hoses should not be allowed to come in contact with oily surfaces and should be

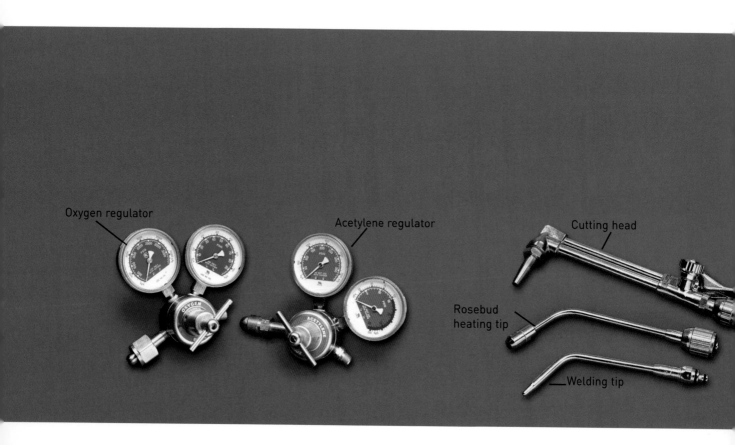

Oxygen regulator

Acetylene regulator

Cutting head

Rosebud heating tip

Welding tip

protected from sunlight and chemical fumes to keep them in good condition. Always drain hoses when you have finished the welding session to remove the pressure, as this will extend the life of the hoses.

Fittings. Fittings connect the hoses to the regulator at one end and the torch at the other. For safety, the nut for the fuel hose (red) uses left-handed threads and has a groove machined around the surface of the nut to help identify that it is for fuel use only. The oxygen (green) hose nut has right-hand threads and no groove. Never interchange the oxygen and acetylene hoses or fittings—they are color coordinated and have specific threads to help maintain consistency and safety. Never force a fitting: brass is a very soft metal, and it is easy to damage the fittings. Always hand-thread and hand-tighten fittings before using a wrench, and be careful not to overtighten them. Never use pliers to tighten fittings, as they will damage the brass fittings.

Check valves & flashback arrestors. Check valves and flashback arrestors are two safety features designed to prevent reverse gas flow or flashbacks. The check valve allows gas to flow from the cylinder to the torch and in that direction only. If the gas pressure within the torch exceeds the hose pressure, a spring closes the valve to prevent backflow. In the event of a flashback, the check valve needs to be replaced. A flashback arrestor is installed between the torch and the hose, and it offers more protection than a check valve. The flashback arrestor prevents burning oxygen from flashing back into the hoses and regulator, which can cause an explosion. The flashback arrestor consists of a check valve, pressure-sensitive valve, stainless-steel filter, and heat-sensitive check valve. In the event of a flashback, the flashback arrestor does not need to be replaced.

When these valves are used, a restriction is created, and gas supply pressure may have to be increased to reduce the chance of fuel starvation—a condition that may cause the torch to overheat.

Torch. The torch mixes and controls the flow of fuel gas and oxygen. The torch consists of valves, a torch body, a mixing chamber, and a welding tip. In a combination torch, the mixing chamber and adjustment valve will be attached to the cutting attachment.

Welding tips. The welding tips attach to the torch body and come in many sizes to create different size flames. The tip size refers to the orifice opening: as the tip size increases, the amount of fuel gas mixture increases and produces a hotter flame. It is important to match the correct size tip to the correct thickness of the metal being welded. Matching the tip size, the welding material, and the gas pressure is critical to creating quality welds.

Oxygen and acetylene hoses with flashback arrestors, torch body, and welding tip

HOW TO SET UP AN OXYACETYLENE OUTFIT

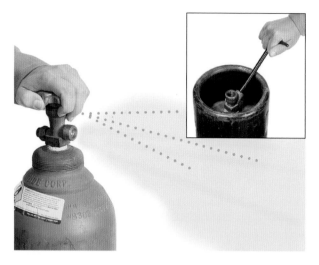

1 Secure the cylinders in an upright position, chained to a cart or strapped to a wall or post. Remove the protective cylinder caps. Wipe off the cylinder valve seats, regulator connections, and hose connections with a clean cloth. Crack open each cylinder valve briefly to expel any trapped dirt particles. "Flat top" acetylene cylinders (inset) may have antifreeze in the recessed valve seat. Use a clean rag to remove the liquid and dry the valve seat. This style acetylene cylinder requires a cylinder wrench to open the valve.

2 Attach the regulators to the cylinders. (The acetylene connectors have left-hand threads.) Always hand tighten, then use a fixed wrench, not a pliers or an adjustable wrench, to tighten. Do not overtighten—a firm seating is all that is necessary. Attach the hoses to the regulators. The acetylene hose is red and left-hand threaded. The oxygen hose is green. *Never* use grease, oil, or pipe dope to lubricate fittings. Grease and oil can ignite spontaneously when they come in contact with oxygen—even without a spark or flame present.

3 Turn the regulator adjustment screws on the oxygen and acetylene regulators counterclockwise until they are loose. (Some regulators have a knob.)

4 Open the oxygen valve slowly all the way while standing to the side in case the regulator gauge glass shatters. Turn the regulator adjustment screw until oxygen begins to flow through the hose, then loosen the regulator adjustment screw to stop the oxygen flow. Slowly turn the acetylene cylinder valve ¾ to 1½ turns.

5 Adjust the acetylene regulator valve until the gas begins to flow, then loosen the regulator adjustment screw to stop the flow. Attach the torch to the hoses. Be sure to pressurize the system and check for leaks before lighting.

HOW TO PRESSURIZE (TURN ON) AN OXYACETYLENE OUTFIT

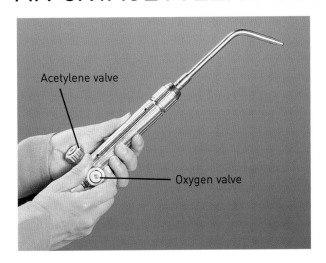

Acetylene valve

Oxygen valve

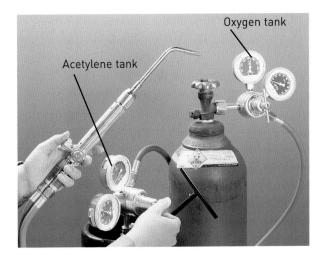

Oxygen tank

Acetylene tank

Make sure both torch valves are off. Turn both regulator adjustment screws counterclockwise until loose. Slowly turn the oxygen cylinder valve on. Once open, turn the valve all the way open to ensure proper seating without leaks. Turn the oxygen regulator adjustment screw clockwise until the gauge reads the desired pressure. (Refer to the manufacturer's specific instructions for operating pressures.) Open the oxygen valve on the torch to check the flowing oxygen pressure. Adjust if necessary and close the torch valve. Slowly open the acetylene cylinder valve ¾ to 1½ turns. (Leave the wrench on the valve if it is a wrench-style valve.) Turn the acetylene regulator pressure adjustment screw until the desired pressure reads on the regulator gauge. Open the acetylene valve on the torch briefly to make sure the flowing pressure matches the desired working pressure. If not, adjust the regulator until the proper pressure is reached.

CHECKING FOR LEAKS

Apply leak-detecting solution to all connections with a small brush. (You can use soap and water so long as the soap is not petroleum based.) If any connections cause bubbles in the solution, tighten the connections and check again.

WARNING

Backfire and flashback are two hazardous situations that can be caused by improper gas pressures. Backfire is the pre-ignition of the acetylene and oxygen inside the tip that causes a popping sound. This may damage the tip or spray molten metal from the weld area. Flashback is the flame burning backward into the torch or hoses, causing a popping or squealing noise. Flashback can cause an explosion in the hoses. Avoid both hazards by matching the tip size to the material being welded and by using the proper pressure settings. Using lower pressures than recommended can cause backfire and possibly flashback.

LIGHTING THE TORCH

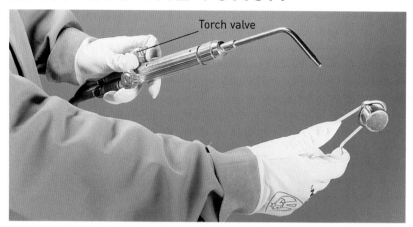

Torch valve

1 Hold the torch in one hand with the thumb and forefinger on the acetylene torch valve. Hold the striker in front of the torch about 3 to 6" away at a slight angle. Turn on the acetylene torch valve ¼ to ½ turn. Immediately use the spark lighter to light the flame.

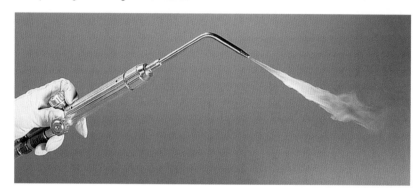

2 The flame will be yellow and smoky.

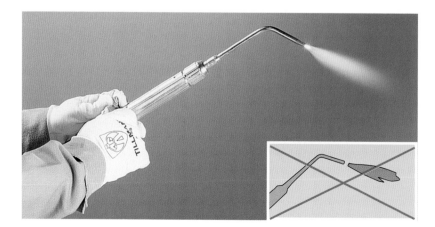

3 Put the striker down and adjust the acetylene torch valve with your right hand so the flame is burning without producing soot. The flame should not be separated from the torch (inset). Open the oxygen torch valve slowly. Adjust the oxygen to get a bright white inner flame and a bluish outer flame. Turn down the acetylene to eliminate the excess acetylene feather if present. When you have finished welding, turn off the oxygen first, then the acetylene.

FLAME & FLAME STATES

The flame of an oxyacetylene torch has two parts—the inner or primary flame and the outer or secondary flame. The flame has different temperatures at different locations. The outside edges are cooler because they are burning with the ambient air, which is only 21% oxygen. The torch tip is cooler because complete combustion hasn't been reached. The hottest area is the tip of the primary or inner flame cone. Here the gases are completely combusted and are insulated by the secondary flame. There are three flame states for the oxyacetylene flame: carburizing, neutral, and oxidizing.

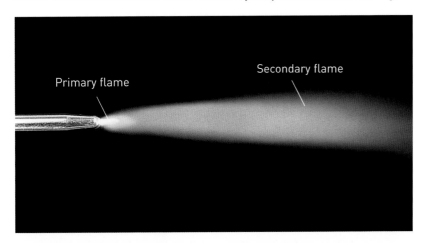

Primary flame

Secondary flame

The carburizing or reducing flame has an excess of fuel. This is a useful flame as it will break down metal oxides to get at the oxygen, thus cleaning the weld area to a small degree. This process adds carbon to welds, which makes them harder. The carburizing flame has a bright white primary flame, an acetylene "feather" around the primary flame, and a bluish white secondary flame with orange edging.

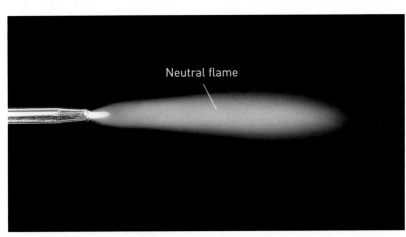

Neutral flame

The neutral flame is the exact point where the feather and the inner cone come together. In this flame, there is exactly enough oxygen present to provide total combustion of the fuel gas. Most welding and cutting operations use a neutral flame. The neutral flame has a bright white primary flame and a colorless to bluish secondary flame.

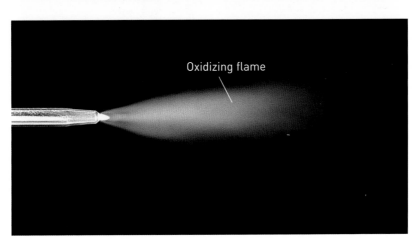

Oxidizing flame

The oxidizing flame has an excess of oxygen. The white cone of this flame is small and pointed and somewhat paler than the neutral flame. A hissing sound often accompanies this flame. This flame is not particularly useful as it hastens oxidizing, which is not desirable in welding. It can, however, be used for removing carbon from molten metal, thus softening the metal.

HOW TO WELD WITH OXYACETYLENE

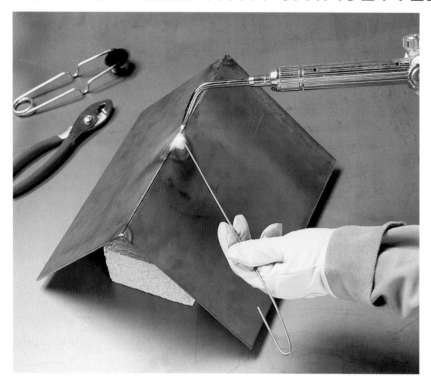

- Check hoses for damage before pressurizing the system.

- Prepare metal for welding by wire brushing or sanding off mill scale and rust. Use acetone or denatured alcohol to remove oil or other chemical residues.

- Use fire bricks to avoid unnecessary heat loss and prevent welding to the welding table.

- See manufacturer's recommendations for appropriate tip sizes and gas pressures.

- Set up materials and clamp if necessary.

NOTE: The directions for oxyacetylene welding are for right-handed welding. Reverse the directions for left-handed welding, or if you find it easier to manipulate the filler rod with your right hand.

1 Select an appropriate filler rod and lay it on the table next to the bricks. Light the torch and adjust to a neutral flame. Pull down your face shield. Place small fusion tack welds at each end of the joint and in the middle if it is a long joint. (A fusion weld uses no filler rod.) Turn off the torch, oxygen first then acetylene, and check that your tacked piece is still in the desired position. If not, use a hammer to move it into position or break the tack weld and reposition.

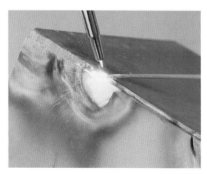

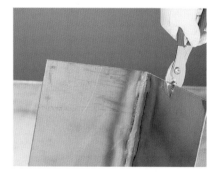

2 With the torch at a 45° angle to the right and oscillating the torch in a ¼ to ½" circle over both metal pieces, create a weld puddle at the right end of your workpiece.

3 When a molten puddle has formed, move the filler rod close to the puddle and flame, but not in it. Begin moving slowly to the left while oscillating and maintaining the molten puddle. Dip the filler rod into the middle of the molten puddle and remove it, but keep it within the heat zone.

4 Continue dipping, oscillating, and moving to the left. As you reach the end of the weld, the cumulative heat buildup may make it necessary to adjust to a shallower angle to deflect heat away from the puddle and prevent burn through. When finished, turn off the oxygen torch valve first, then the acetylene torch valve. The weld should penetrate to the back without burning through.

DEPRESSURIZING AN OXYACETYLENE OUTFIT

When you have finished welding, or if you are going to stop welding for more than 10 or 15 minutes, depressurize your setup. The hoses and regulators are designed to leak small amounts of fuel and oxygen if they are pressurized and not being used, so it is important for safety and economy to depressurize them. You will soon be so comfortable with the pressurizing and depressurizing steps that it won't seem like an inconvenience at all. It is important to do the fuel gas and oxygen in separate steps, to prevent having mixed, unburned oxyacetylene in the torch.

1. Close the fuel gas cylinder valve.
2. Open the fuel gas torch valve until both gauges read zero.
3. Loosen (counterclockwise) the fuel gas regulator adjustment screw.
4. Close the fuel gas torch valve.
5. Close the oxygen cylinder valve.
6. Open the oxygen torch valve until both gauges read zero.
7. Loosen (counterclockwise) the oxygen regulator adjustment screw.
8. Close the oxygen torch valve.

OXYFUEL TIPS

GAS OPTIONS:

The most useful fuel gas for oxyfuel welding is acetylene. Its neutral flame generates the highest heat and heat concentration with the lowest chemical interactions with the molten metal. Other fuel gases, such as MPS or MAPP and propylene can be used for brazing and cutting applications. Natural gas and propane can be used for brazing and heating operations. Make sure that the equipment—hoses, regulators, torches, and tips—are designed for use with the gas you choose.

GAS	DENSITY (AIR = 1)	TEMPERATURE OF NEUTRAL FLAME WITH OXYGEN
Acetylene	0.906	5589° F
Methylacetylene-propadiene (MPS or MAPP)	1.48	5300° F
Natural gas	0.62	4600° F
Propane	1.52	4579° F
Propylene	1.48	5250° F

Maintenance. The primary maintenance for oxyacetylene welding equipment is cleaning the torch tip. A dirty tip will spark, pop, and often direct the flame sideways. Tip cleaners are inexpensive sets of cleaning rods used to clean the various sized tip orifices (see photo, page 43). The set usually comes with a small file to smooth the tip. To clean, simply insert the correctly sized tip cleaning rod into the orifice, and pull back and forth. Do not rotate the tip cleaner as this may enlarge the orifice. Also, do not use drill bits to clean tips. The sharp edges of the bit will cut grooves into the tip. Cleaning rods have smoothed edges and are sized especially for tips. Do not clean tips while the torch is lit.

BRAZING

Brazing is very similar to soldering since flux is applied to tightly fitted metal parts that are then heated to the point where filler material will melt and be drawn into the joint. Silver soldering and hard soldering are terms incorrectly used to refer to brazing. Brazing is different from soldering because it takes place at temperatures over 840°F and below the melting point of the base metals. The metals are not fused but held together by the filler metal adhering to the base metals through capillary action.

There are a number of industrial brazing processes, such as dip brazing, furnace brazing, and induction brazing. The home welder is likely only to do torch brazing. Torch brazing can be done with an oxyfuel torch using acetylene as the fuel gas, or any of the other fuel gases (see page 49). *NOTE: Different fuel gases require different regulators, hoses, torches, and tips.*

For brazing to work, the gap between the parts must be between 0.002 and 0.010". If the gap is too tight, the flux and filler will not flow evenly through the joint. If the gap is too big, the strength of the joint is lessened. Gaps between parts can be measured with a feeler tool, available at automotive stores. Items to be brazed must be absolutely clean and free from rust, corrosion, grease, oil, and cleaning compound residues.

Brazing is often used commercially to join dissimilar metals such as tungsten carbide saw teeth to a steel saw blade. Another common use for brazing is in lugged bicycle frames. Because nearly every metal can be joined using brazing, it is highly suitable for art applications.

Brazing supplies include flux and silver solder (above). Braze welding requires flux-coated rods or separate flux and rods (sold individually or in bulk).

Thoroughly clean and flux both sides of the joint area. Using a small torch tip, heat the entire joint area until the flux turns clear and starts to run. Add enough filler metal to fill the joint. (Silver alloy is shown.) After the metal has cooled, the flux residue can be removed with hot water.

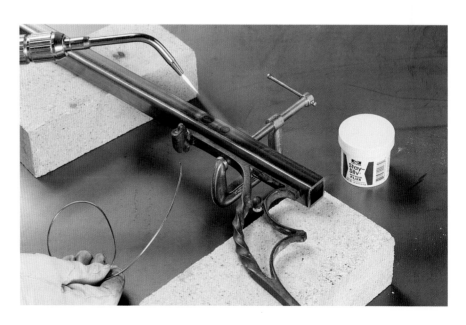

BRAZE WELDING

Braze welding is similar to standard oxyacetylene gas welding except the parent or base metals are not melted, so there is no molten puddle. Instead of a steel alloy filler rod, a flux coated brass filler rod is used. Braze welding is often incorrectly referred to as brazing. Braze welding does not use capillary action to pull filler material into the joint—the filler metal is deposited as fillet or groove welds.

The brazing rod is melted by the heat of the metal and the flame, but it should not be held in the flame itself. The parts for braze welding should fit tightly, but the gap is not as important as with brazing.

Braze welding is used for joining dissimilar metals and for metals of different thicknesses. This technique is often used to repair cracked or broken cast iron.

Braze welding has less distortion than oxyacetylene welding because less heat is applied to the parts. A disadvantage is that it is not as strong as welding where the base metal is melted, but a well-made braze weld is still sufficient for most non-structural applications. Because the base metals do not need to be melted, braze welding can be done with any of the fuel gases listed on page 49.

If you are creating a piece that will be welded and braze welded, you must be careful to complete the non-brazed welds first. The heat involved with all other welding processes will boil off the brass alloy of a braze weld, ruining the weld and creating toxic fumes.

1 Braze welding is useful for joining thin metals, such as this expanded metal, to thicker metals. Heat both parts, directing more heat toward the thicker part. It may take a long time for the thicker metal to heat. Using fire bricks will prevent a metal tabletop from absorbing any heat.

2 When both parts glow a dull cherry red, touch the flux coated rod to the joint. The flux and the filler metal will melt. If the metal is molten or the fluxed rod comes in contact with the flame, the flux will burn and the filler metal will boil. This results in a poor joint in addition to giving off toxic fumes.

ELECTRICITY FOR WELDING

Arc welding and cutting processes, including shielded metal arc welding (SMAW), gas metal arc welding (GMAW), gas tungsten arc welding (GTAW), and plasma arc cutting (PAC), all use electricity to generate the necessary heat. Understanding electricity is not necessary to use these processes, but having a basic understanding of electrical terminology and concepts will help you understand how to better utilize the welding arc.

Electric current is the flow of electrons through a conductor from a high concentration of electrons (negative charge) to a low concentration of electrons (positive charge). As the electricity flows through a conductor, it generates an amount of heat based on how much resistance the conductor offers to the flow. Common conductors utilized for electrical circuits are copper and brass. An arc is simply a sustained electrical discharge across an air gap, commonly called the *arc* or *arc length*. Because air is highly resistant to the flow of electrons, the movement of electrons across the air gap generates a lot of heat. The heat generated by an arc can be as high as 11,000 degrees Fahrenheit, but half of that heat is dissipated and does not reach the molten base metal. Here are some other important terms to understand when it comes to electricity:

- **Amperage/Amps (A):** The volume of electrons flowing through a conductor. The amperage controls the size of the arc.

- **Electrical units:** The units used to refer to electricity are voltage, amperage, and wattage.

- **Open-circuit & operating voltage.** The open-circuit voltage is the voltage that exists at the electrode tip when the machine is on but no arc has been struck. The higher the open-circuit voltage, the easier it is to strike an arc. Open-circuit voltage is typically limited between 50 and 80 volts—any higher and the chance of electrical shock increase. The operating voltage is the voltage when the arc is struck and the circuit is complete, allowing the flow of electrons. This is usually between 17 and 32 volts, depending on the arc length, the type of electrode, and the polarity selected.

- **Voltage/Volts (V):** The measure of electromotive force, pressure, or electric potential. The voltage controls the size of the air gap that the arc can cross. The higher the voltage, the larger the gap the arc can cross.

- **Wattage (W):** The measurement of the amount of electrical energy or power contained in the arc. Wattage affects the width and depth of a weld.

- **Welding power:** Power for welding may be supplied in one of the following ways:

  - **Constant current (CC):** As the amperage fluctuates, the voltage increases or decreases to keep the total power relatively the same.

  - **Constant voltage (CV):** The arc voltage is maintained even when the current (amperage) changes. This means if the arc length is changed, which changes the current flow, the voltage stays relatively the same.

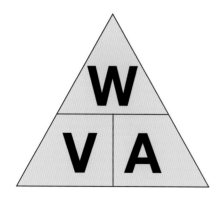

Volts, amps, and watts are related to each other numerically: volts × amps = watts.

POWER SOURCES
FOR WELDING

Three sources produce the low-voltage, high-amperage combination that arc welding needs:

• **A mechanical generator** produces power through the use of a gasoline or diesel engine.

• **A step-down transformer** takes available high-voltage alternating current and changes it to a low-voltage, high-amperage current.

• **An inverter** uses solid-state electronics to change the current without the weight of a transformer.

CURRENTS USED
FOR WELDING

Alternating current (AC) is standard available household current. The electron flow changes direction two times per cycle, or 120 times per second, giving the standard 60-hertz cycle common in the United States. Welding with AC means the electrode and the work piece alternate between positive and negative, so the welding heat is distributed evenly to both. This balances penetration and buildup.

Direct current (DC) electrons flow in one direction only. A rectifier is used to convert alternating current to direct current. Direct current has two polarity options:

• **Direct current electrode negative (DCEN),** formally called *straight polarity*, means the electrode is negative and the work piece is positive—electrons are flowing from the electrode to the work piece and the heat is concentrated on the work piece. Approximately ⅔ of the heat is concentrated on the work piece, allowing for greater penetration. This also can cause distortion and a very unmanageable puddle.

• **Direct current electrode positive (DCEP)**, formally called *reverse polarity*, means the electrode is positive and the work piece is negative—the electrons are flowing from the work piece to the electrode, concentrating approximately two-thirds of the heat on the electrode. This, however, produces the best welding characteristics, as we have better control of the puddle and distortion caused by excessive heat.

DUTY CYCLE

The duty cycle is described as the amount of continuous arc time, or running time, at a given power output for a welding machine, recorded as a percentage of time. Because welding machines produce heat internally, they need time to cool. A 60-percent duty cycle means that a machine can run continuously at that setting for six minutes, and then it will need to cool for four minutes. Duty cycle is critical for construction and industrial welding equipment and application as the machines are utilized heavily in a production environment. For a home or hobby shop, duty cycle should still be researched, but most machines sold today have a high enough duty cycle to maintain the workflow.

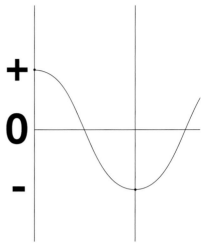

Alternating current alternates between positive and negative polarities, passing through a zero, or no current, point between the two.

NOTE

Work leads and work clamps are sometimes referred to as ground leads and ground clamps. This is not technically correct, as the electricity is not being grounded. Instead, the work lead and clamp are allowing for completion of the circuit back to the machine.

SHIELDED METAL ARC WELDING (SMAW)

Shielded metal arc welding (SMAW) is also referred to as *arc* or *stick welding*. The process involves the heating of the base metal to fusion or welding temperature by an electric arc that is initiated when the covered electrode is tapped or scratch started on the base metal. The coating or flux covering on the electrode provides both flux and shielding gas for the molten weld (the *puddle*). The flux breaks down to create shielding gases to keep the atmospheric air away from the molten weld pool, and it also creates a protective slag that forms over the top of the weld to protect it as it solidifies or cools. The electrodes come in 9- to 15-inch straight lengths, in a range of wire thickness from $1/16$ inch to $3/8$ inch, hence the name *stick*. SMAW is used extensively for fabrication, construction, and repair work, because the machinery is inexpensive and fairly simple, and the electrodes are fairly inexpensive. However, SMAW does not work well for thin metals, less than $1/8$ inch. Electrodes need to be changed frequently as they are consumed, and the protective slag coating must be chipped off each weld.

SAFETY

SMAW uses electricity, so there is always the possibility of receiving an electric shock or electrocution. When an electrode (stick) is placed into the electrode holder, it is live. If the electrode touches anything that the work clamp is in contact with, the circuit will be completed and the arc will ignite. To prevent this from happening, always remove the electrode from the holder when you are not actively welding. Always handle electrodes with dry gloves, remember electricity and water do not mix well, and do not weld while standing in water or on a damp floor. The electrode will be hot after welding, so take care when you dispose of the electrode stubs. A metal bucket with sand for snuffing out the stubs is a good addition to the welding shop. SMAW produces ultraviolet and infrared rays, harmful fumes, and hot spatter. Protect your eyes with a #10 to #14 filter in a full-face helmet or hood. Heavy-duty leather welding gloves and a welding jacket with leather

Power source

Electrode holder

Work clamp

A shielded metal arc welder.

sleeves are necessary to protect you from molten spatter. Proper ventilation from an exhaust hood or fan is important to protect the welder from the fume cloud produced while welding. Also, it is a good idea to screen off your welding area so others are protected from the intense light of the arc.

EQUIPMENT

SMAW machines are available as either alternating current (AC) or direct current (DC), or with the capability of switching between the two. The machine itself is simple, in that it merely converts high-voltage, low-amperage line current into low-voltage, high-amperage welding current. Output is controlled by one knob. You will often hear a smaller SMAW machine referred to as a "buzz box."

AC welding machines meet most home and small shop needs, are inexpensive, and are readily available. Because the alternating current cycles through a zero current between the positive and negative polarities, it can be difficult to strike and maintain an arc. DC machines are easier to use, and have many home and hobby applications, but they are more expensive. Because DC current can have its flow reversed, a DC machine has more versatility in terms of the electrodes that can be used. This allows for a wider range of welding positions, metal thicknesses, and metal types that can be welded. This versatility makes a DC shielded-metal arc welder well worth the extra expense.

The equipment itself consists of the welding machine (which usually has one adjustment knob), an electrode lead with electrode holder (sometimes called a stinger, whip, or lead), and a work lead with work clamp. You will often see work leads and work clamps referred to as ground leads and ground clamps. The work lead and clamp are not grounding the electricity; they are simply completing the circuit back to the machine.

SAFETY

+ Welding helmet with #10 to #14 filter
+ Leather, wool, or cotton long pants; leather jacket
+ Heavy-duty welding gloves
+ Safety glasses
+ Hat
+ Leather boots or shoes
+ Ventilation

Useful tools for shielded metal arc welding are a pliers, chipping hammer, and wire brush.

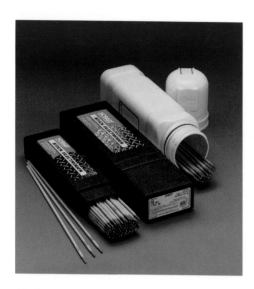

SMAW electrodes are sold in 5- or 10-pound boxes. Specialty electrodes may be bought a pound at a time. It is important to keep electrodes dry either in the original box or a storage container.

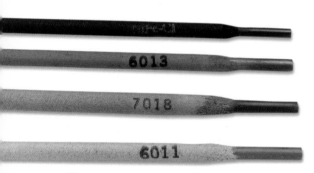

Electrodes for SMAW are stamped with a numeric code so you can tell which to use with arc power and which are for DC.

Electrodes. SMAW electrodes are solid, round, metal wires coated with flux and other components. In addition to producing shielding gas and flux, the covering may also contain additional metals for filler or alloying elements in the weld.

The American Welding Society (AWS) publishes standards for the electrodes. Electrodes come in diameters ranging from $\frac{1}{16}$ to $\frac{3}{8}$ inches in increments of $\frac{1}{32}$ inches. The electrode diameter measures the wire itself, not the diameter of the wire including the covering. The electrode designation is inked onto the covering near the bare end of the electrode. The number classification for SMAW electrodes begins with the letter E, denoting electrode. The first two or three numbers to the right of the letter E denote tensile strength. For example, in the designation *E7018*, a four digit number, the first two numbers to the right, 70, represent 70,000 psi tensile strength. The tensile strength of a properly formed weld should withstand a force of 70,000 pounds per square inch. In another example, *E11018*, a five digit number, the first three numbers to the right, 110, represent 110,000 psi tensile strength. The digit to the right of tensile strength denotes the welding positions recommended for this electrode: 1 = all, 2 = flat grooves and flat or horizontal fillet welds. The last digit, when accompanied by the digit to its left (the digit for position), designates the recommended current and polarity for the electrode and any notes. We can also add suffixes that denote alloys that have been added to the electrode. Each electrode manufacturer may have a number of electrodes of a specific AWS designation that are slightly different and have been tailored to a specific use.

Use the manufacturer's guidelines when selecting an amperage to use with each specific electrode. The guidelines are developed based on the size of the electrode and the thickness of the material to be welded. If the amperage range cannot be found, a good rule of thumb for a $\frac{1}{8}$" diameter electrode is 90 to 120 amps. Adjust the amperage to match your skill level and preferred travel speed.

The most commonly used electrodes for SMAW include E6010, E6011, E6013, and E7018. The E6010 is designed to work on DCEP and is not a fast-freeze rod used to fuse root joints. The E6011 and E6013 will work with DC and AC current, and E6013 on DC is the easiest electrode for a new welder to use in terms of striking an arc and maintaining a consistent arc. The E7018 electrode is best suited for DCEP, and is a medium-penetrating weld with excellent strength. To make the proper electrode selection, talk with your welding supplier about what type of welder you are using, the type of welds you are making, and the metal with which you are working.

Slag. Shielded metal arc welding produces a weld that is covered by a protective slag that is formed during the welding process. This ceramic-like coating helps clean the weld zone and float any impurities to the top of the weld. The slag also controls the cooling of the weld zone to allow the weld to solidify in an oxygen-free atmosphere. This slag must be chipped or scraped away before the weld bead can be inspected and before another weld pass is placed on top or any finish or paint is applied. Safety glasses and face shield should always be worn when removing slag.

HOW TO WELD WITH SHIELDED METAL ARC

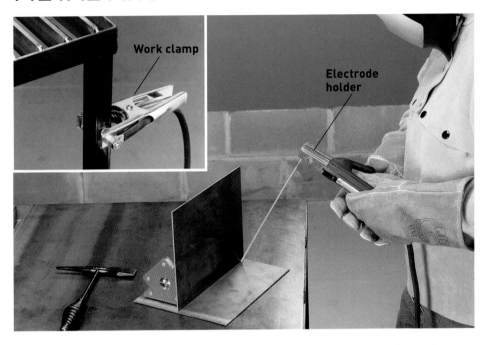

Work clamp

Electrode holder

PRECUTTING CHECKLIST

- Electrode and work leads are tightly attached to machine terminals.

- Cable insulation is free of cuts, damage, and wear.

- Electrode lead is tightly connected to electrode holder.

- Work lead is tightly connected to clamp.

- Work clamp is firmly clamped to worktable or workpiece.

- Floor is dry.

- Ventilation is adequate to draw fumes away from welder's face.

1 Set up your material to be welded. Make sure the electrode holder is not touching the workpiece or worktable. Attach the work clamp to the table (inset) or workpiece. Turn on the machine. Adjust the range switch for the desired amperage. Wearing leather gloves, place an electrode in the electrode holder. Position the electrode over the area to be tacked, flip down your helmet, and tap or scratch the electrode on the area to be tacked to strike an arc. After making your tack welds, remove the electrode from the electrode holder. Check to see that the tacked pieces are aligned properly. If not, use a hammer to move them into alignment or break the tack welds and retack. Chip slag from tack welds so it does not contaminate the final weld.

2 Replace the electrode in the electrode holder, and position the electrode over the left side of the area to be welded. Hold the electrode at a 10 to 20° angle to the right. Flip down your hood and scratch or tap to strike an arc. The distance between the metal and electrode should not exceed the thickness of the electrode bare wire diameter. Move slowly to the right until the weld is completed.

3 To remove slag, hold the workpiece with pliers at an angle and scrape or knock the slag with the flat blade of the chipping hammer. Wear safety glasses when chipping slag.

GAS METAL ARC WELDING (GMAW/MIG)

Gas metal arc welding (GMAW), also referred to as MIG (metal inert gas) welding or wire feed, is a process whereby a consumable electrode (wire) is automatically fed through a welding torch or whip, along with a continuous flow of shielding gas supplied through an external cylinder.

With GMAW, four different process modes transfer metal from the electrode to the work piece. Choosing the correct mode depends on the welding process, the welding power supply, and the consumable—each mode has unique characteristics and applications.

Several variables dictate the type of transfer you use, including the amount and type of welding current, electrode surface, electrode diameter, shielding gas, and the contact tip-to-work distance. Transfer mode also affects your choice of filler metal used.

SHORT-CIRCUIT TRANSFER

In this method, the electrode short circuits upon touching the work, causing the metal to transfer as a result of the short. This happens at a rate of between 20 and more than 200 times per second. One advantage of short-circuit transfer is its low energy. Short-circuit transfer is usually used on thin material ¼ inch or less, as well as on root passes on pipe that has no backing. It is suitable for welding in all positions and generally calls for smaller-diameter electrodes: 0.023, 0.030, 0.035, 0.040, and 0.045 inches.

A gas metal arc welding setup consists of a power source, wire feed (inside power source), work cable with clamp, supply cable with gun, and regulator.

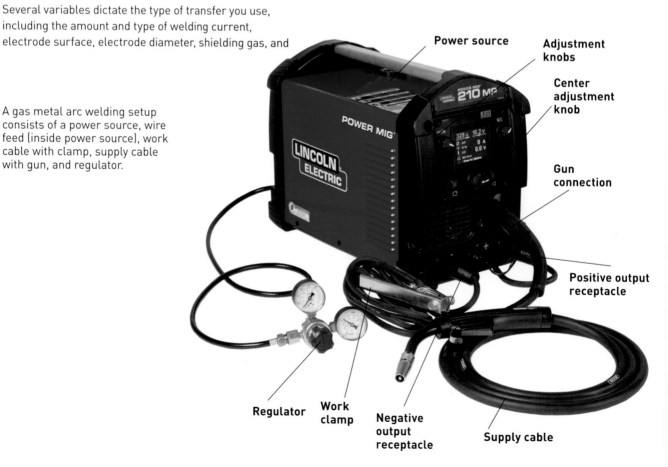

Power source

Adjustment knobs

Center adjustment knob

Gun connection

Positive output receptacle

Supply cable

Negative output receptacle

Work clamp

Regulator

GLOBULAR TRANSFER

In globular transfer, the weld metal transfers across the arc in relatively large droplets—usually greater than the diameter of the electrode being used. This mode of transfer generally is used on carbon steel only and uses 100 percent CO_2 shielding gas. It is often used to weld in the flat and horizontal positions, because the large droplet size is more difficult to control if used in the vertical and overhead positions—at least when compared to the short-circuit arc transfer. This mode generates the most spatter.

SPRAY TRANSFER MODE

Spray transfer is called that because of the spray of tiny molten droplets across the arc. Spray transfer usually is narrower than the diameter of the wire, and it uses fairly high wire speeds and voltage and amperage. Unlike short-circuit transfer, with spray transfer, the arc is on at all times once it is established. Spray transfer produces little spatter, making it a good choice on thick metals in the flat and horizontal positions.

PULSE-SPRAY TRANSFER

In this mode, the power supply cycles back and forth between a high spray transfer current and a low background current. This cycling allows for cooling of the weld pool during the background cycle—a major difference from a true spray transfer. In the perfect scenario, one droplet in each cycle transfers from the electrode to the weld pool. Because of the low background current, pulse-spray transfer is ideal for welding out of position on thick sections, since it uses higher energy than that provided by short-circuit transfer—thus producing a higher average current with better side-wall fusion. And it also can be used to lower heat input and reduce distortion when high travel speeds are either not needed or cannot be achieved because of limitations in equipment or throughput.

The GMAW process has many advantages. The gun or torch can be held a uniform distance from the weld, unlike shielded metal arc welding where the distance from the electrode holder to the weld becomes shorter as the electrode is consumed. With the power control

The wire feeding mechanism consists of a spindle to hold the wire spool, drive wheels, rollers, and a tension adjustment. Most welders now have a reference chart with recommended settings.

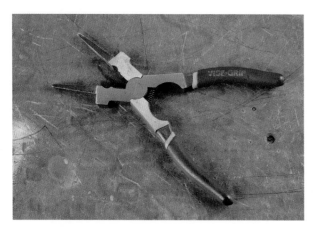

A pair of MIG pliers is a handy tool for gas metal arc welding. These are designed for hammering, wire cutting, insulation brushing removal and installation, and drawing out wire. They are also ideal for removing spatter from inside, outside, and the nozzle end.

Most home GMAW welders use 1-pound or 10-pound wire spools. Larger spools are generally more economical.

trigger built into the torch or gun, a welder can be in place and ready to weld without accidentally striking an arc. When ready, the welder flips down the helmet and pulls the trigger without being off target. Because the electrode is the filler, the welder does not have to coordinate the filler rod and torch in opposing hands. The GMAW torch can be held steady with both hands to complete a uniform consistent bead. As the shielding gas is supplied from an external cylinder through the torch, the GMAW welds a smooth, uniform, and free-from-slag weld—no need to chip or grind. The GMAW process operates at a lower amperage and is a relatively cool welding process, so 22- and 24-gauge metal can be welded with little to no distortion. The process also requires narrower beveling for thicker plate welds, so less time is needed for weld groove preparation.

Some disadvantages of GMAW include the fact that the gas shielding nozzle does not allow for welding in tight spaces without some modifications. And the shielding gas can be disrupted or blown away from the welding zone by drafts or wind areas, making this process unsuitable for working outdoors.

EQUIPMENT

Power supply. In the small GMAW units, the power supply and feeder unit are integrated into the same cabinet or case. In larger multipurpose machines, the feeder unit is separate from the power supply, allowing for more versatility. The power supply converts the standard AC into DC. AC is not appropriate for GMAW because the constant change from positive to negative does not provide a steady, consistent arc. The welder may adjust the voltage, polarity, and wire feed rate (usually given as

IPM or inches per minute), with the amperage determined by the wire feed rate and the contact tip–to-work distance. When using the power source to weld with a shielding gas, the machine should be set to DCEP—that is, so the electrode is positive and the work piece is negative. If used without a shielding gas with flux core wire, the machine should be set to DCEN—that is, so the electrode is negative and the work piece is positive. In smaller machines, this is typically accomplished by switching wires inside the machine between two terminals marked + and -. On larger multipurpose machines, this can be accomplished by simply flipping a switch, or the machine may change automatically as the welding mode is changed. Always check the manufacturer's recommended settings for selecting polarity for a given electrode.

Wire feed unit. The wire feed unit consists of a drive motor, drive rolls, spool adapter, and spool of electrode (wire). The wire speed is adjusted via a knob and is adjusted for the metal thickness being welded. The drive rolls are marked with a diameter, and should match the diameter of the wire. Most machines come with a set of drive rolls that are two sided for multiple diameters of wire.

Torch/gun. The GMAW gun, sometimes referred to as a torch, is attached to the power supply via a power cable that carries the electrical power, the power control for the switch, a liner for the filler metal or wire to be fed through, and a liner for shielding gas. The power control switch, when depressed, initiates the electrical current needed to energize the filler metal (wire), start the flow of shielding gas, and start the drive motor to push the wire through the torch. The gun generally has a gooseneck shape, although a straight neck and flexible neck guns are

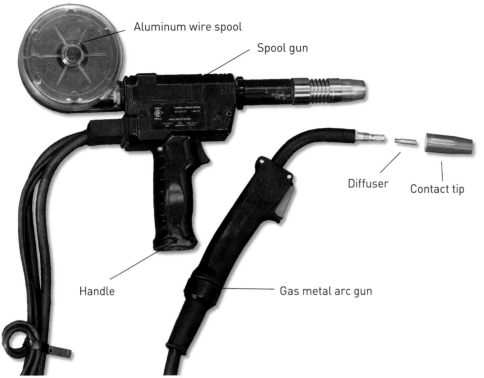

Aluminum wire spool

Spool gun

Diffuser

Contact tip

Handle

Gas metal arc gun

A spool gun is used for feeding aluminum wire because it often breaks or misfeeds through a standard cable. A standard gas metal arc gun consists of a handle, gas diffuser, contact tip, and nozzle.

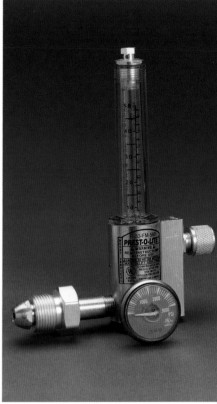

A flow meter registers gas flow with a small floating ball.

also available. A contact tip is screwed into a gas diffuser attached to the gooseneck. The pair serves as the contact point for the wire to become electrified and disseminate the gas into the shielding nozzle. The contact tip orifice or opening must be the same as the diameter of the wire to ensure the wire is not allowed to arc or short circuit.

Work cable/work clamp. This is attached to the work piece, or to a metal surface that the work piece is placed on, to complete the circuit.

Electrodes. When choosing an electrode (wire), consider the composition properties, cleanliness of the metal, and the type of shielding gas that will be used. If you will be welding out of position, this will also be a factor, as certain arc stabilizers can be added to the chemical components added to the solid wire. There are literally dozens of choices for GMAW filler metal: consult your welding supplier for the best choice for your application.

Electrodes are labeled with alphanumeric codes that describe their type, tensile strength, whether the wire is solid or tubular (tubular is flux cored), and chemical composition. For mild steel, ER70S-6 is a good general-purpose wire. In this example, the E represents *electrode*, the R represents *rod*, the 70 represents *70,000 psi tensile strength*, the S represents a *solid wire*, and the suffix, 6, represents a specific chemical composition added to the

filler metal. GMAW filler metal is also available for welding aluminum and stainless steel. Available wire sizes are 0.023, 0.030, 0.035, and 0.045 inches.

Using a wire feed machine without a shielding gas requires the use of a flux-cored wire often referred to as a core wire, and the process is then called flux-cored arc welding (FCAW). Think of this filler metal as a hollow tube with the flux added to the inside. This still makes it possible to energize the exterior of the wire as it is fed through the contact tip.

Shielding gas. The shielding gas for GMAW is supplied by an external cylinder and flow meter or regulator and connects to the power supply with a hose designated for inert gas. Pure carbon dioxide (100 percent) is suitable for general GMAW. It provides an aggressive, deep, penetrating arc, and is inexpensive compared to other gases. One disadvantage to 100 percent CO_2 is that leaves an excessive amount of spatter, which can require a lot of post-weld cleanup. Carbon dioxide can be mixed with argon in a 75 percent argon/25 percent carbon dioxide mix. It is slightly more expensive than pure carbon dioxide but yields welds with less spatter and acceptable penetration. Pure argon is used as a shielding gas in gas tungsten arc welding (GTAW); if used on GMAW, it produces a shallow penetrating weld.

HOW TO SET UP THE WIRE FEED

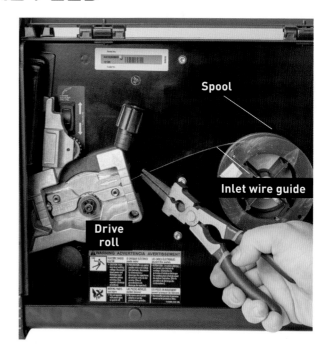

Spool

Inlet wire guide

Drive roll

1 Contact tubes are stamped with their size. Make sure the contact tube and the drive roll grooves are the correct size for the electrode you are using. Place the wire spool on the spindle, and secure it with the pin lock, lock ring, or wing nut. Make sure it is feeding in the proper direction.

2 After releasing and cutting the crimped wire end, hold the wire firmly with a pliers. The wire is tensioned, and the entire spool will rapidly unroll if you do not hold it firmly. Swing the tension arm or pressure roll out of the way. Push the wire through the inlet wire guide, through the groove on the drive roll, and out through the outlet wire guide.

NOTE

Unless your machine has an inch switch for advancing wire or a purge switch for activating gas flow, you will use the gun trigger to accomplish both tasks. Using the gun trigger means the wire is "live" and will arc to anything the work clamp is touching.

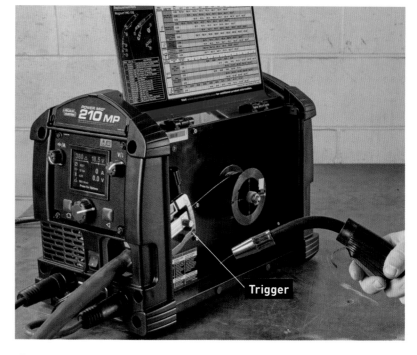

Trigger

3 Swing the tension arm or pressure roll back into position. Adjust tension according to manufacturer's directions. On a new machine, the drive roll will already be set. Make sure the wire is aligned perfectly straight—not up and down or side to side—on the drive roll. Turn the machine on, turn the wire speed to its highest setting, pull the cable straight, and depress the trigger. Some machines have an inch button that feeds the wire without supplying power to the contact tip or wasting shielding gas while loading wire.

HOW TO SET UP THE SHIELDING GAS FLOW METER

1 Open the cylinder valve briefly to clear out any dirt. Wipe the cylinder threads with a clean, dry cloth.

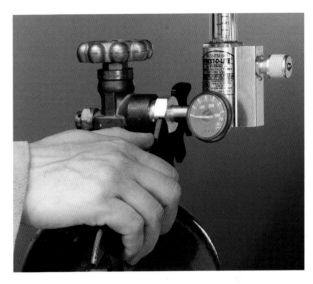

2 Attach the flow meter to the cylinder. Use a fixed wrench to prevent damaging the brass fittings.

Flow meter

3 Turn the knob on the flow meter clockwise to tighten it so no gas can flow through it. If the flow meter is open when you open the cylinder valve, the high pressure can damage the flow meter.

4 Attach the hose to the flow meter and to the welding machine. Slowly open the cylinder valve, then, once it is open, open the cylinder valve completely. To set the flow rate, turn on the machine. Depress the trigger to activate the gas flow, and turn the flow meter knob counterclockwise. While gas is flowing, continue turning the knob until the meter registers the proper flow rate. Make sure that the flow meter or flow gauge is rated for use with the shielding gas you are using. For example, an argon/carbon dioxide flow meter cannot handle pure carbon dioxide.

HOW TO WELD WITH GAS METAL ARC

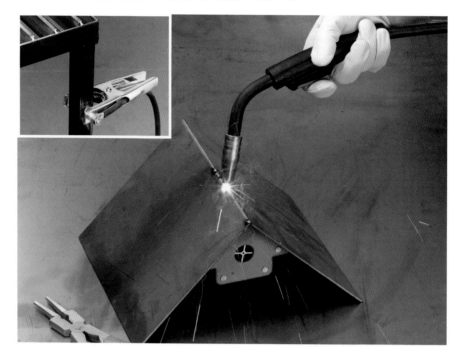

1 Prepare and fit up the material to be welded. Attach the work cable clamp to your welding table (inset) or the workpiece. Turn on the machine and shielding gas. Adjust the wire speed and voltage according to the manufacturer's recommendations. Cut the electrode to ⅜" stickout. Put the tip of the electrode at the point of the first tack weld, flip down your helmet, and pull the trigger. Place tack welds evenly around the weld area. Trim the electrode for proper stickout at the end of each use. Check to see that the workpiece is still aligned properly. If not, adjust with a hammer or break the tack welds and retack.

2 Start at the left end of the weld. Hold the gun with both hands and position it at a 20° angle with the tip pointing to the left. Put the tip of the electrode at the beginning of the weld, flip down your helmet, and pull the trigger.

3 Begin welding, moving steadily to the right. GMAW has a distinctive sizzling sound. Popping and snapping indicate dirty material or an improper voltage or wire speed adjustment.

4 A finished GMAW is smooth with even ripples or weave pattern, no slag, and little spatter.

POSTWELDING SEQUENCE

When you have finished welding with a GMAW rig, it is important to shut it down properly so you are set for your next welding session. If you do not turn off your shielding gas, you will lose much of it to bleed off even if you weld the next day.

1. Close the gas cylinder valve.

2. Press the gun trigger or purge switch until the flow meter or flow gauge reads zero.

3. Turn off the flow meter.

4. Turn off the machine.

5. Coil the supply and work cables and store them off the floor.

FLUX CORED ARC WELDING

Flux cored arc welding uses the same wire feed and power supply as GMAW but usually without the gas shielding. This is convenient for welding outdoors in windy conditions, but the weld is not as clean. To use flux cored wires, most welding machines need to be converted. Usually this involves changing the output polarity and installing a gasless nozzle. It also might require installing proper drive rolls and a different cable liner. Otherwise, follow the same steps to weld with flux cored wire as with solid wire.

Flux cored wire welds are covered with a layer of slag.

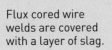

"Stub out" is when the electrode welds to the base metal without melting or breaking off. This is caused by the voltage being too low, the wire feed being set too fast, or holding the gun too close to the work when starting. Correct the settings, grind off the stub outs, and restart the weld.

Stub out results in short sections of wire poking out of the weld.

Use dippable anti-spatter gel to prevent spatter buildup on the nozzle and contact tip. Simply dip the hot nozzle into the gel.

MAINTENANCE & TROUBLESHOOTING

A small amount of maintenance on your GMAW equipment will result in better, more consistent welds and longer-lasting equipment. Spatter—tiny pieces of the electrode or base metal that have sizzled off—builds up on the nozzle and the contact tip. Remove this spatter frequently as you weld so it does not interfere with gas flow or electrical conductivity. Turn off the machine, remove the nozzle, and use the closed point of the MIG pliers (page 60) to ream out the nozzle and file the end of the contact tip. An anti-spatter gel is available in which to dip the hot nozzle and contact tip. Anti-spatter spray can be used to coat the contact tip and the nozzle. This spray is also useful for coating your welding table to prevent spatter from sticking to your work surface. Eventually you will need to replace the nozzle, but regular maintenance and proper welding techniques will extend the life of a nozzle considerably.

Contact tubes become worn because they are a soft copper alloy and the electrode is steel. The electrode will wear through the contact tip, possibly making the electrical flow irregular. Contact tips should be visually inspected for wear on a regular basis. The orifice on a worn contact tip will appear oval instead of round. Tips generally are good for about eight hours of continuous welding use. *NOTE: Always turn off the welder before changing the contact tip.*

It is important to keep the welder fan motor and rectifier free of dirt and dust to prevent it from overheating. If your welder is stored in a dusty shop, or you use it infrequently, keep it covered. Use low pressure air to blow dust and dirt from these assemblies. Use a vacuum to remove dirt from the wire feed mechanism.

The supply cable needs to be cleaned occasionally as well. Check the manufacturer's recommendations for frequency, but it's generally recommended to clean after you've used 50 pounds of flux cored wire or 300 pounds of solid wire. With the power off, remove the cable from the machine. Remove the gas nozzle and contact tip from the gun, lay out the cable straight, and use low-pressure air to blow into the gun end. (Using high-pressure air may create a dirt plug that will clog the cable permanently.)

Always store the supply and work cables and the gas hose off the floor and away from chemicals, hot sparks, and sunlight.

REMOVING A BIRD'S NEST

A bird's nest is a tangle of wire in the wire drive mechanism. This happens when the wire drive rolls continue feeding wire but the cable is blocked, the outfeed tube is misaligned, the wire is stubbing out on the base metal, or the wire has been welded to the contact tip. The drive rolls are set to slip in these cases, but if the tension has been adjusted too tightly, they may continue to push wire.

Wear safety glasses when fixing a bird's nest. To fix a bird's nest, turn off the machine and cut the wire as it comes off the spool. Remember the wire is under tension, so you must hold the spool end, then tie it off through one of the spool holes. The tangled wire also may be storing tension, so it may fly out of the drive roll area. Release the tension arm or roller arm. Pull the wire out of the drive mechanism and the supply cable.

GMAW TROUBLESHOOTING

PROBLEM	SOLUTION
Wire feeds, but no arc can be struck	• Check to see that the work clamp is connected.
Weld bead is full of holes like a sponge (porosity)	• Make sure the shielding gas is on. • Adjust shielding gas to a higher flow rate. • Eliminate drafts. • Verify correct polarity.
Weld bead is tall and looks like a rope	• Make sure the voltage setting is appropriate for the material's thickness. • Reduce your travel speed.
Excessive spatter or dirty welds	• Reduce the gun angle. • Hold the nozzle closer to the work. • Decrease the voltage.
Arc starts and stops	• Check the wire feed for steady feeding. • Check the cable connections. • Clean the weld materials.

GAS TUNGSTEN ARC WELDING (GTAW/TIG)

Gas tungsten arc welding (GTAW), commonly called *TIG* (*tungsten inert gas*) *welding*, and sometimes referred to as Heli-Arc (the L-TECH trade name) because early uses of TIG welding used helium as a shielding gas, is a process that generates an arc between a nonconsumable tungsten electrode and the work piece. A shielding gas protects the electrode and the weld, and filler metal may or may not be used.

Gas tungsten arc welding differs from other arc processes because the electrode is nonconsumable and not used as a filler material. GTAW is more like oxyacetylene welding in terms of the skills needed to manipulate a torch with one hand and a filler rod with the other. GTAW requires another layer of coordination, because most machines also use a foot-activated amperage control. Like GMAW, GTAW is a clean process, because the shielding gas eliminates the need for flux and resultant slag.

SAFETY

Because GTAW is such a clean process, welders are often tempted to weld without gloves or in a short-sleeve shirt. This is not recommended. As it is an arc process, the arc produces ultraviolet light at a higher level than other processes, and because there are no fumes or smoke, those light rays are unfiltered and can cause severe burns. It is critical to cover all exposed skin to prevent UV burns. Filter requirements for eye protection are a minimum of a shade #10, and if you have an auto-darkening hood, be sure it is rated for GTAW. Some entry-level auto-darkening helmets are not designed for GTAW.

EQUIPMENT

The basic equipment needed for GTAW is a constant-current welding machine, cable with torch, work cable and clamp, electrode, and inert gas cylinder with regulator and flow meter. Optional equipment includes a remote amperage control and a water-cooled torch. Although a midrange shielded metal arc welding machine can be used to deliver the current for GTAW, a dedicated good-quality GTAW machine delivering the current as AC or

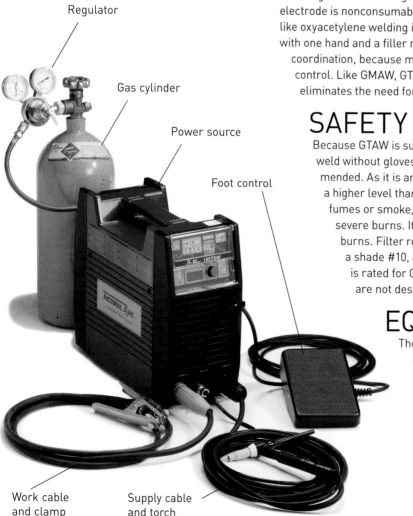

A gas tungsten arc welding setup.

Regulator

Gas cylinder

Power source

Foot control

Work cable and clamp

Supply cable and torch

DC provides an optional high-frequency output for no-touch arc starting, has a remote control option for a foot-pedal control, and has a solenoid for shielding gas control. The combination of AC and high frequency makes it possible to weld aluminum with good results. The newest inverter-based GTAW machines have advanced current control capabilities and are becoming more affordable for the home or hobby shop.

Torch & cables. The GTAW torch holds the electrode and delivers the shielding gas. It can be air- or water-cooled. The torch parts include a cup or nozzle, collet body, collet, end cap, and torch body. The collet and collet body hold the electrode firmly in place to allow for the transfer of electricity to generate the arc. The electrode diameter is typically $\frac{1}{16}$", $\frac{3}{32}$", or $\frac{1}{8}$", and the collet body and collet must match this size to ensure a tight connection so the electrode cannot arc in the collet. The cup or nozzle directs the shielding gas to protect the electrode, puddle, and filler metal from atmospheric air to ensure a clean, quality weld. Air-cooled torches typically come in sizes for use with less than 200 amps—more than 200 amps will require the use of a water-cooled torch.

Nozzles, or *cups*, provide a controlled amount of shielding gas to cover the weld pool, determined by their size, which can range from $\frac{1}{4}$" to $\frac{3}{4}$" in diameter. A smaller nozzle provides less coverage than a larger nozzle. Nozzles also vary in length, short to extra-long, and also in their price and performance.

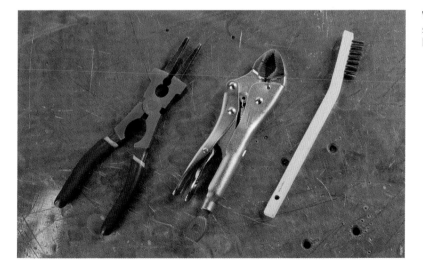

Welpers, locking pliers, and a stainless-steel wire brush are handy for GTAW.

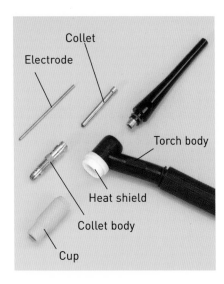

Parts of a gas tungsten arc welding torch.

Labels in image: Collet, Electrode, Torch body, Heat shield, Collet body, Cup

The most cost-effective nozzles are 90 or 95 percent alumina oxide—these are adequate for lower amperage applications. On higher-amperage applications, however, they do not resist thermal shock very well, and in this use they may deteriorate, crack, and fall off.

Lava nozzles cost more than alumina oxide, but they are also more resistant to cracking. They work well for medium amperage applications, but because they have varying wall thicknesses around the inside diameter, gas coverage may be unequal.

SHIELDING GASES

Argon. Because it is an inert gas, argon does not react with other compounds or elements. It is about 1.4 times heavier than air. The inert properties of argon make it ideal as a shield against atmospheric contamination, which is why it is used in many welding processes. Because its potential for ionization is low, argon promotes good arc-starting characteristics and arc stability.

Helium. Because of its high thermal conductivity and potential for high ionization, helium is a good choice for a shielding gas when increased heat input is sought and when there is a low tolerance for oxidizing elements, such as when welding aluminum and magnesium.

Gas flow rate. Gas flow rate can range from 10 cubic feet per hour (CFH) to more than 60 CFH, depending on the current developed, torch size, shielding gas composition, welding position, and operating current—not to mention the surrounding work environment.

As a rule, a higher operating current requires a larger torch nozzle and higher gas flow rates. Operating currents greater than 150 amps also require that you use a water-cooled GTAW torch. This not only controls the heat buildup, but also can allow you to use a smaller tungsten diameter, thereby reducing user fatigue. Gas density—the weight of the gas relative to air—influences the minimum flow rate required to shield the weld adequately.

Because argon is about 1.4 times as heavy as air and 10 times as heavy as helium, the gas flow rates must be increased to maintain quality when working in vertical or overhead positions. On the other hand, helium can be more effective than argon when working overhead, because it floats. When working in the flat position with helium-enhanced blends, maintaining weld quality requires that you increase gas flow when compared to using argon alone. Flow can be 50 percent or more than with pure argon.

ELECTRODES

There are six common tungsten electrodes available for use in GTAW, and choosing the right one is a crucial first step. Tip preparation is also critical. The electrode choices are the following: pure tungsten, 2 percent thoriated, 2 percent ceriated, 1.5 percent lanthanated, zirconiated, and rare earth. The end preparations include balled, pointed, and truncated.

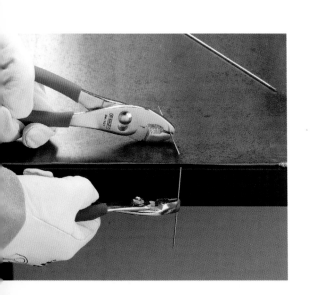

Tungsten is very brittle, so electrodes can be scored with a file, then snapped over a sharp table edge.

Tungsten is a rare metallic element used in the creation of GTAW electrodes. Tungsten's hardness and high temperature resistance facilitate the transfer of the welding current to the arc. Tungsten has the highest melting point compared to other metals, at 3,410 degrees Celsius.

Whether pure tungsten or an alloy of tungsten and other rare-earth elements and oxides, these nonconsumable electrodes come in a variety of sizes and lengths. The proper electrode choice depends on the base material type and thickness and on whether you are using an AC or DC welding process. Choosing between balled, pointed, or truncated end preparations also is crucial in optimizing your results.

Color coding eliminates confusion over electrode types. The color appears at the tip of the electrode.

PREPARING THE ELECTRODE

Prior to use, the cut end of the electrode must be sharpened to a point or melted to a ball. The tip may be ground to a point or chemically sharpened. The electrodes come in 7-inch lengths. To increase the number of points available, score the electrode with a file or cut-off wheel, and snap it in half. Tungsten is very hard but brittle, so it is easy to grasp each end of the electrode with pliers and snap it in half over a sharp table edge.

Because all tungsten electrodes look and feel the same regardless of their composition, it is important to keep them clearly separated by type. The color codes will wear

To safely hold tungsten electrodes for sharpening with a grinder, insert them into the chuck of a drill. Running the drill while sharpening the tungsten ensures a uniform point. Move the tungsten across the wheel to prevent grooves in the wheel edge.

off, or, if you point each end of your electrode, be ground off. It is helpful to have clearly labeled containers for each type of electrode.

Two critical factors in grinding the electrodes are the grinding wheel and the grinding direction. You must use a hard, fine grinding wheel dedicated exclusively to tungsten. Metal particles left on the wheel from grinding aluminum or steel would contaminate the tungsten,

ELECTRODE CHARACTERISTICS

ALLOY	AC/DC	TRAITS & USES	COLOR
Pure tungsten	AC	• Welding of aluminum • Forms ball at tip	Green
1% Thorium oxide (thoriated)	Primarily DC, but also AC	• Easier arc starts • Carries more current • Low-level radioactive	Yellow
2% Thorium oxide (thoriated)	DC	• Long life • Easier arc starts	Red
Zirconium oxide (zirconiated)	AC	• Similar to pure tungsten but carries more current	Brown
2% Cerium oxide (ceriated)	DC	• Not as good as thoriated but not radioactive	Orange
Lanthanum oxide (lanthanated)	DC	• Similar to ceriated	Black

Chemical sharpeners are available for sharpening tungsten.

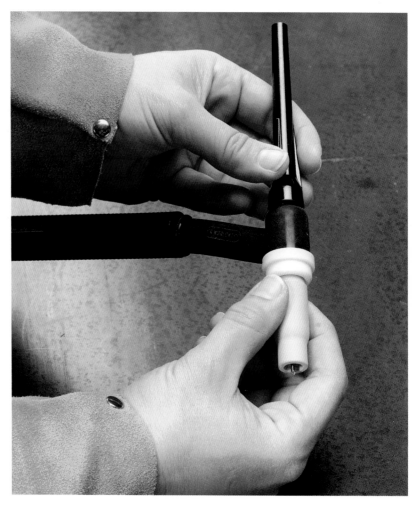

Tightening the back cap locks the tungsten electrode into place. The tip of the electrode should extend beyond the end of the cup by a distance of three times the electrode's diameter.

which causes erratic arc behavior and poor weld quality. An extremely hard material, tungsten will become hot as it is ground. Sharpen the electrode tip so that grinding marks run lengthwise down the tip, not in a circular or crosswise pattern. Lengthwise grinding focuses the electron flow toward the tip; circular grinding causes the arc to be unfocused and possibly jump sidewise from the electrode rather than off the tip point. Chemical means also can be used to sharpen tungsten by dipping a hot tungsten rod into a chemical agent. The length of the taper on the tungsten tip should be two to three times the diameter of the tungsten.

ANATOMY OF THE GTAW MACHINE

Gas outlet

Power input

Settings button

Process mode button/reset

Encoder knob

Power switch

Negative output

Remote control socket

Positive output

The basic setup, however,
is the same. Dedicated GTAW machines, such as
this one and the one pictured on
page 68, have many advanced
features.

SETTING UP THE GTAW MACHINE

Dedicated GTAW machines, such as this one and the one pictured on page 68, have many advanced features. The basic setup, however, is the same.

The positive and negative output receptacles are for the torch and work clamp connections.

The gas outlet connects to the torch gas hose.

The remote control socket is for the foot or finger control. Using a remote control allows the welder to control the amount of current while welding.

The operating mode selects high frequency or scratch starts.

The process mode button/reset selects either AC or DC.

AC balance control using the settings button and the encoder knob allows for adjusting AC power to be more positive or more negative to create either more cleaning action or deeper penetration. Advanced machines can adjust the AC power to be as much as 90% positive or negative, rather than the 50% for standard AC power.

Start current controls the current for arc starting.

Welding current sets the range for welding. The foot or finger control then operates within this range.

Slope down time gradually reduces power to the arc without extinguishing the arc. This allows for the weld crater to fill before the arc is extinguished.

Gas post flow controls how long the shielding gas will flow after the arc is extinguished.

PREWELDING CHECKLIST

- If using a water-cooled torch, check for leaking.

- Check all cables for wear and damage.

- Clean and fit up parts to be welded.

STRIKING AN ARC

Unless you have a high frequency option, you will need to physically strike an arc—called a scratch start. Rest the cup on the workpiece at a sharp angle. Move the tip until it briefly contacts the work, then angle it back again to start the arc. After the arc is started, lift the cup off the workpiece and establish the proper torch angle.

High frequency allows the arc to jump the gap without needing to create physical contact between the electrode and the workpiece.

HOW TO WELD WITH GTAW

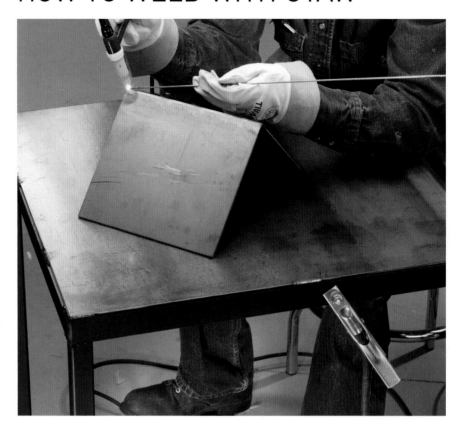

1 Set the controls based on manufacturer's recommendations for the material to be welded. Turn on the machine, and turn on the water pump, if available. Attach the work clamp to the welding table or workpiece. Flip down your helmet, activate the foot or finger control if using one, and strike an arc by scratching the tip of the tungsten against the base metal. If your welder has a high frequency option, you do not need to scratch start the arc. Place a tack weld at each end of the joint to be welded. You may be able to tack the joint by simply fusing the two pieces with the heat of the torch, or you may have to use filler rod.

2 Position yourself to weld from right to left (if you are right handed) with the torch at a 15° angle to the right of center. Hold the filler rod in your left hand. Position yourself so you can comfortably hold the torch and filler rod for the duration of the weld.

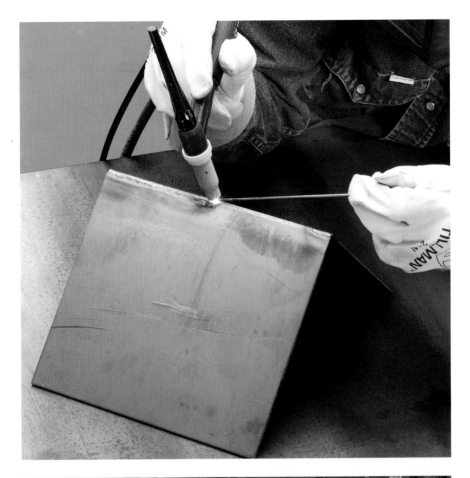

POSTWELDING SEQUENCE

- Turn off the cylinder valve.
- Purge gas from the gas line.
- Turn off the flow meter or gauge.
- Turn off the machine.
- Coil hoses and cables off the floor.

3 When a molten puddle has formed, dip the tip of the filler rod into the middle of the molten puddle. Keep the filler rod at a low angle to prevent disturbing the shielding gas. Keep the tip of the filler rod near—but not in—the puddle. Move the electrode to the left and continue the melting and dipping process.

4 As you approach the end of your weld, you may need to adjust your travel speed, because the buildup of heat in the material makes the molten puddle form more quickly at the end of the weld than at the beginning. You may also need to adjust the torch angle to be shallower (not shown here) so that less heat is directed into the base metal.

TROUBLESHOOTING

Filler metal for GTAW comes in rod form and ranges from 1/16 to 3/16 inch in size. Rods are available in a variety of alloys, including aluminum, chromium and chromium nickel, copper, nickel and nickel alloys, magnesium, titanium, and zirconium. Specific alloy compositions are available for creating specific weld types on specific base metals. These filler metals are similar to those used in oxyfuel welding, with the exception of the carbon steel rods, which are not copper coated as they are for oxyfuel.

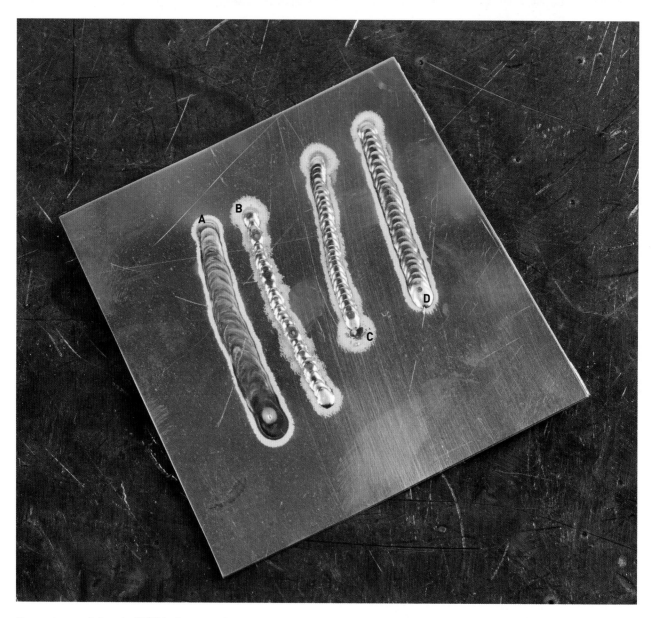

Becoming proficient in GTAW takes practice, and identifying problem welds is an important step. Weld A is too hot. Increase the travel speed or decrease the amperage. Weld B is too cold and is simply sitting on top of the base metal rather than penetrating it. Decrease the travel speed or increase the amperage. Weld C was done too quickly. Travel speed needs to be controlled and consistent. Weld D is a good-quality weld with even ripples, good penetration, and a moderate crown.

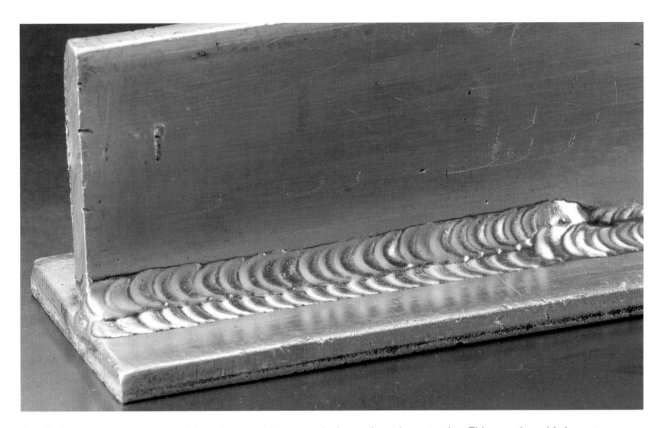

A well-done gas tungsten arc weld on aluminum has even ripples and good penetration. This sample weld shows two passes to create a fillet weld on ¼" stock.

GTAW Troubleshooting

PROBLEM	SOLUTIONS
Weld looks porous or sooty	• Make sure shielding gas is on and is correct type. • Make sure shielding gas cylinder is not empty. • Eliminate drafts. • Make sure base metal is totally dry. • Clean base metal thoroughly. • Increase gas flow rate.
Base metal distorts	• Tack weld parts before welding. • Clamp parts down to rigid surface. • Scatter welds to diminish heat buildup.
Unstable arc	• Adjust electrode to work angle. • Clean base metal thoroughly. • Clean electrode. • Connect work clamp to workpiece. • Bring arc closer to work.
Electrode is rapidly consumed	• Make sure polarity and current settings are correct. • Increase electrode size. • Increase gas flow. • Decrease current. • Increase gas postflow time. • Use proper shielding gas.

WELDING PROJECTS

Here's a chance to put all these welding and cutting techniques to use. The following chapters have directions for creating 34 different projects. Each contains a detailed cutting list, technical drawing, and step-by-step directions. Among the shop projects you will find a sturdy welding table and a handy welding cart with cyinder racket angled top. Included in the home décor projects are a delightful wine rack and a handy solution to creating a coffee table out of a slab of stone or wood. The outdoor life projects will spark your creativity with an add-your-own-found-objects gate.

SHOP

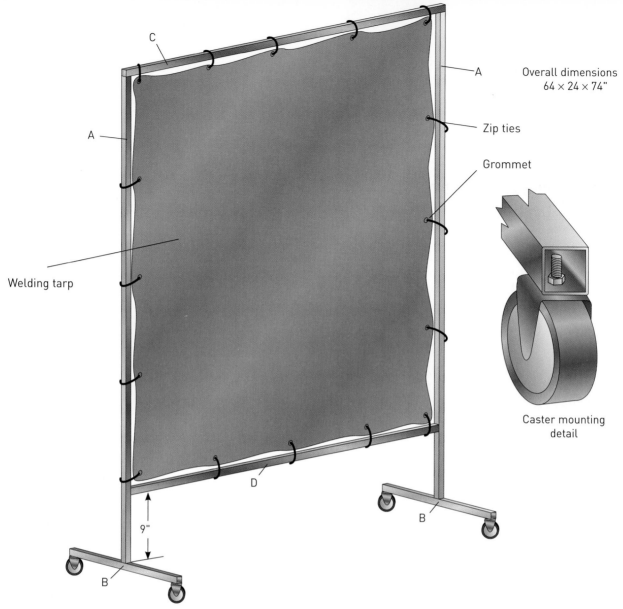

Overall dimensions
64 × 24 × 74"

Zip ties

Grommet

Welding tarp

Caster mounting
detail

ROLLING WELDING CURTAIN

For arc welding and plasma cutting, it is very important to screen your work area so other shop workers, passersby, family members, and pets are not exposed to the damaging rays. This rolling curtain is quick and easy to make, and it offers the necessary screening while you work. You can purchase ready-made welding tarps in a variety of colors, shapes, and sizes (see Resources, page 236). Or you can make your own welding tarp with 12-ounce cotton duck, fire retardant, and grommets.

PART	NAME	DIMENSIONS	QUANTITY
A	Sides	1 × 1" square tube × 72"	2
B	Wheel supports	1 × 1" square tube × 24"	2
C	Top crossbar	1 × 1" square tube × 64"	1
D	Bottom crossbar	1 × 1" square tube × 62"	1

Sixteen gauge or thin wall tube is sufficient for this project.

HOW TO BUILD
A ROLLING
WELDING CURTAIN

ATTACH THE WHEEL SUPPORTS
TO THE SIDES

1. Clean all parts with denatured alcohol, acetone, or degreaser. Prepare weld joint areas by wire brushing until shiny.
2. Cut the sides and wheel supports (A and B) to length.
3. Clamp one side piece to the work surface. Center a wheel support at the end of the side piece to form a T.
4. Check for square and tack weld along the butt joint between the two parts. Turn the assembly over and reclamp it to the work surface. Check that the wheel support is still square to the side piece. Make a final weld along the butt joint.
5. Repeat steps 3 and 4 to assemble the second side piece and wheel support.

WELD THE WHEEL SUPPORT T-JOINTS

1. Clamp a side and wheel support assembly to the work surface so the wheel support hangs over the edge.
2. Weld the T-joint between the wheel support and the side piece. Turn the assembly over and weld the second T-joint.
3. Repeat steps 1 and 2 for the second side and wheel support.

INSTALL THE CROSSBARS

1. Cut the top and bottom crossbars (C and D) to length.
2. Place the top crossbar over the side pieces (see photo). Check for square, clamp in place, and tack weld.
3. Make a mark 10" up from the bottom of each wheel support. Align the lower edge of the bottom crossbar with the marks, check for square, and tack in place.
4. Flip the assembly over and check for square by measuring across both diagonals. If the measurements are equal, the structure is square. Clamp the assembly in place and complete the welds. Flip the assembly over and finish the welds on the other side.

APPLY FINISHING TOUCHES

1. Paint the framework, if desired.
2. Drill holes for the threaded post swivels ½" from the end of the wheel supports. Install the casters.
3. Attach the welding tarp to the framework with zip ties.

MATERIALS

- 1 × 1" square tube (27')
- 4 threaded swivel casters, at least one locking
- 5 × 5' welding tarp
- Zip ties

On a steel table, finish weld the wheel supports to the sides, and then weld the top and bottom crossbars to the sides.

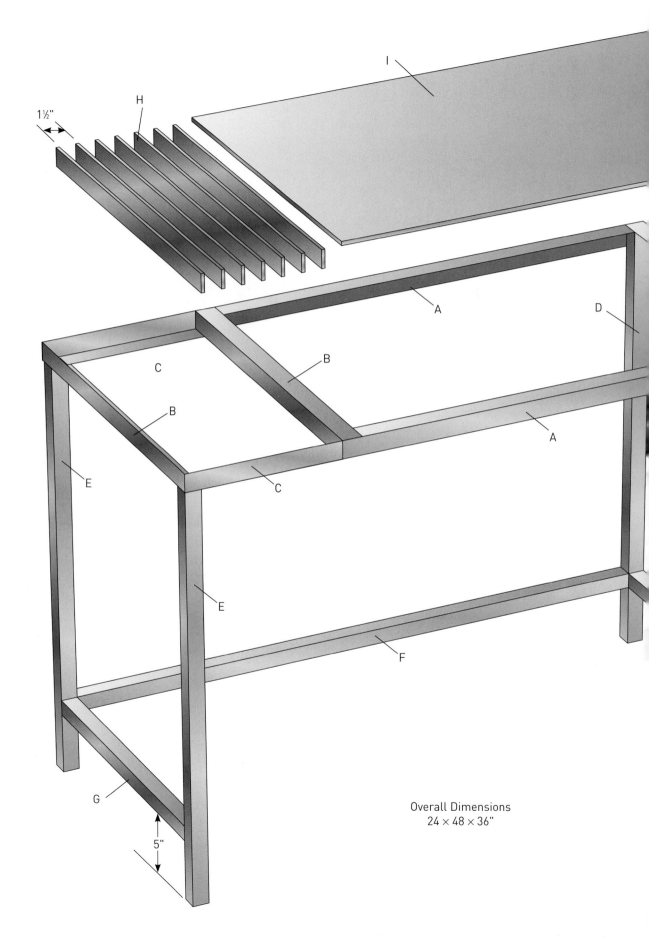

H

1½"

I

A

D

B

C

B

A

C

E

B

E

F

G

5"

Overall Dimensions
24 × 48 × 36"

WELDING TABLE

A sturdy welding table is the heart of any metal shop. Used for arc welding, cutting, or oxyfuel welding, the welding table is a versatile addition to your shop. The model shown here is sized to allow you to work while sitting on a stool or while standing. You may want to check the scrap bin at the local steel yard to see if you can find a bargain price for the tabletop. Some yards sell plate in $4 \times 2'$ pre-cut sections that you can cut to size. Or you can have the piece custom cut. You can use material that's thinner than $3/16"$, but thicker material is better as the tabletop will have less distortion from the welding heat and the heat generated when you grind spatter off of it. Set the sheet metal for the tabletop on top of sawhorses or a workbench for use as a work surface to build the rest of the table.

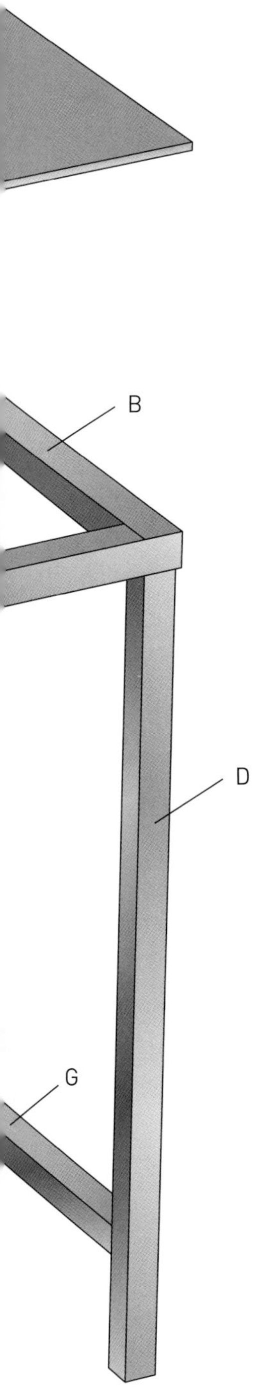

MATERIALS

- $1/8 \times 1^1/2 \times 1^1/2"$ angle iron (12')
- $1/8 \times 1^1/4 \times 1^1/4"$ square tube (20')
- $1/4 \times 1^1/2"$ flat bar (14')
- $3/16"$ sheet metal (2 × 2')
- 4 leg levelers

PART	NAME	DIMENSIONS	QUANTITY
A	Tabletop supports (front & back)	$1/8 \times 1^1/2 \times 1^1/2"$ angle iron × 36"	2
B	Tabletop supports (sides)	$1/8 \times 1^1/2 \times 1^1/2"$ angle iron × 24"	3
C	Cutting table supports (front & back)	$1/8 \times 1^1/2 \times 1^1/2"$ angle iron × 12"	2
D	Right side legs	$1/8 \times 1^1/4 \times 1^1/4"$ square tube × 36"	2
E	Left side legs	$1/8 \times 1^1/4 \times 1^1/4"$ square tube × 34$^1/2$"	2
F	Stretcher (rear)	$1/8 \times 1^1/4 \times 1^1/4"$ square tube × 45$^1/2$"	1
G	Stretcher (sides)	$1/8 \times 1^1/4 \times 1^1/4"$ square tube × 21$^1/2$"	2
H	Cutting table slat	$1/4 \times 1^1/2"$ flat bar × 23$^7/8$"	7
I	Tabletop	$3/16"$ rolled steel 24 × 36"	1

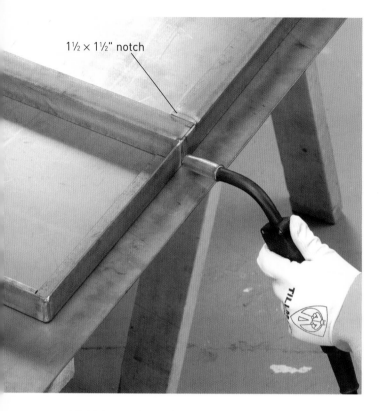

1½ × 1½" notch

Weld the cutting table section to the tabletop assembly. The angle iron flange for the tabletop will face up to support the tabletop. The angle iron flange for the cutting table will face down to form a well to support the cutting table strips.

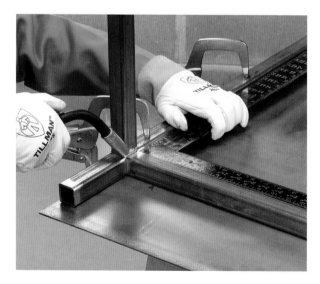

Place the back stretcher between the right and left side leg assemblies. Make sure the stretcher is square to the legs, then weld in place.

HOW TO BUILD A WELDING TABLE

WELD THE TABLETOP SUPPORTS

Cutting notches for the side supports in the front and back tabletop supports is easier than cutting 45° miters, and provides more welding surface area.

1. Cut the tabletop supports (A and B) and cutting table supports (C) to length. Cut 1½" notches where each end of the three side supports (B) fits against the tabletop supports to create a 90° angle joint.

2. Place the tabletop front support (A) and a tabletop side support (B) together to form a right angle. Check for square and tack weld.

3. Repeat step 2 using the tabletop back support and another side support.

4. Assemble these two right angles to make a square. Check all corners for square, and check the assembly for square by measuring both diagonals— they should be equal. If not, adjust the supports so the assembly is square.

5. Complete each outside corner weld, re-checking for square as you go. Flip the assembly over and complete the remaining welds of the joints.

COMPLETE THE TABLETOP SUPPORTS

The cutting table supports are positioned with the flange at the bottom to hold the ¼"-thick strips. The tabletop supports are positioned with the flange at the top to support the tabletop (see photo, above left).

1. Place the remaining side support (B) and the front and back cutting table supports (C) at right angles to form three sides of a rectangle. Check the corners for square, and tack weld the pieces together.

2. Assemble the pieces so the tabletop supports have the flange at the top, the cutting table supports have the flange at the bottom, and the cutting table supports are abutting the tabletop supports. Tack weld where the tabletop sides butt together. Turn the assembly over and weld the remaining joints.

PREPARE THE LEG ASSEMBLY

1. Cut the legs (D & E) to length.

2. Measure the completed tabletop. It may be slightly more or less than 24" deep and 48" wide. Adjust the stretcher lengths (F & G) to those measurements, minus the 1¼" for each leg thickness. Cut the stretchers to length.

Tack weld cutting table slats to the angle iron every 1½" to form the cutting tabletop.

3. Mark both sets of side legs 5" up from the bottom. Starting with the right side legs (D), place a side stretcher (G) between the legs with the bottom of the stretcher aligned with the 5" mark.

4. Align the stretcher at a 90° angle to the side legs and clamp the assembly to your work surface. Tack weld the top inside angle at each end of the stretcher.

5. Repeat this process for the left side legs.

FINISH THE LEG ASSEMBLY

1. Place the leg assemblies on their sides with the rear side down, and clamp to your work surface.

2. Position the rear stretcher (F) between the right and left leg assemblies, aligning the bottom of the stretcher with the 5" mark.

3. Align the stretcher at a 90° angle to the assemblies, and clamp in place. Tack weld the inside angles (see photo, bottom previous page).

ASSEMBLE THE TABLE

The right legs fit inside the corner made by the angle iron, while the left legs are set back ⅛" from the edges of the angle iron.

1. Turn the tabletop and cutting table assembly upside down, then set the leg assembly into it.

2. Clamp the pieces in place. Check for square on both sides, front, and back.

3. Tack weld all corners. Check for square again, then weld all pieces into place.

INSTALL THE CUTTING TABLETOP

Tack welding the cutting table grating allows you to remove and replace the ¼" strips as they become worn from the cutting torch.

1. Cut the cutting table slats (H) to length.

2. Place a slat onto the ledge formed by the cutting tabletop supports 1½" from the tabletop edge.

3. Tack weld the top edge of the slat to the angle iron.

4. Place another slat 1½" from the first slat, and tack weld in place. Continue building the cutting tabletop in this manner until complete (see photo, above).

5. Grind down the welds on the top of the tabletop support assembly.

FINISH THE TABLE

1. Place the tabletop (I) onto the assembly, and tack weld twice on each side.

2. Weld the tabletop to the supports using 1" or 2" weld beads at both sides of each corner and twice along each side.

3. Grind down the welds, if desired. Wire brush, sand, or sandblast the entire table.

4. Paint the table, but do not paint the tabletop or cutting grate.

5. Install leg levelers as needed.

WELDING CART WITH CYLINDER RACK & ANGLED TOP

The ability to easily access your welding equipment can increase efficiency and the ease with which the machine can be used. As one of your first projects in the shop, fabricate a machine cart with an angled top for ease of use. Also use a rubber mat beneath the welder and insulated chain to secure the gas cylinder—both can be found at your local hardware store or home center.

A versatile welding cart can accommodate either a GMAW or SMAW/GTAW machine.

MATERIALS

- 1 × 1" 11-gauge steel square tubing (three 12' lengths)
- ½" solid round stock CRS (21")
- 10-gauge HRS sheet (14 × 18")
- 14-gauge CRS sheet (27 × 25½")
- Wheels (2)
- Pins
- Insulated chain (18")
- Hooks (2)

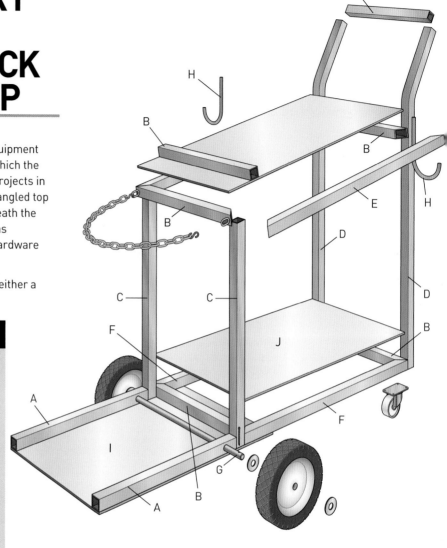

PART	DIMENSIONS	STOCK	QUANTITY
A	15"	1 × 1 steel square tubing	2
B	14"	1 × 1 steel square tubing	6
C	20"	1 × 1 steel square tubing	2
D	36"	1 × 1 steel square tubing	2
E	24"	1 × 1 steel square tubing	2
F	26"	1 × 1 steel square tubing	2
G	21"	½" solid round stock CRS	1
H	10"	⅜" solid round stock CRS	2
I	14 × 18"	10-gauge HRS sheet	1
J	13½ × 23½"	14-gauge CRS sheet	1
K	13½ × 25½"	14-gauge CRS sheet	1

HOW TO BUILD A WELDING CART WITH CYLINDER RACK & ANGLED TOP

FABRICATING THE FRAMES

1. Cut parts B, E, and F for this step. Make sure the bevels cut on each end are opposing, meaning the short points of the cut should both face in toward the center of the tube. Bevel each end at 45°. (This can be accomplished with an oxyacetylene torch, plasma cutter, or cut-off wheel.) For part D only, drill a ½" diameter hole at each end, located 1" in from the ends and centered on the tube.
2. Using two pieces of part B and two of part E, lay out the frame for the lower shelf, forming a 14 × 26" rectangle, with the beveled corners forming a picture-frame shape.
3. Check the assembly to make sure it is square by measuring the diagonals. If the frame is square, both measurements will be the same. Adjust if necessary until the frame is perfectly square.
4. When the frame is set, tack each corner in two places, check again, and then weld each corner all around if the measurements are correct. (To make sure the frame remains square during welding, a temporary diagonal brace may be tacked across the frame to hold it in place. Remove the brace after the assembly has cooled, and then grind off the tacks.)
5. Repeat the above process with two pieces of part B and two of part F to form a 14 × 26" upper frame.

PREPARE THE VERTICAL UPRIGHTS

1. Cut two pieces of part C, and then bevel one end of each piece to 10°. Then cut two pieces of part D, leaving the ends square.
2. Clean the cuts, and then bend each tube (D) to form a slight radius. If you don't have a tubing bender, you can bend the tubing easily by making a series of ⅛" bird's-mouth cuts: measure from one end of the tubing and place a mark at 27" (point A). From point A, measure and place five marks spaced every ½" from point A. At each mark, cut a ⅛" bird's mouth.
3. Weld each joint, and grind smooth. You have now created a rolled handle upright. Repeat for the second upright using the first one as a template to ensure the two match.

Tack part J to the lower frame, between the uprights.

ASSEMBLE THE FRAME

1. Tack the vertical uprights to the lower frame (14 × 24" frame). Check that the uprights are parallel and square to the frame.
2. Tack part J to the lower frame, between the uprights.
3. Tack the upper frame to the vertical uprights, checking to make sure the frames are square.
4. Tack one part B to the top of the upper frame 3" from the top of the 20" upright.
5. Tack part I to the upper frame. Check for squareness, and tack temporary diagonal braces where necessary for welding.
6. Place part B between the ends of the curved uprights, and tack in place.
7. Stitch-weld the shelves, and weld the frame corners all around.
8. Tack part H to the lower side of the lower frame.
9. Tack parts A to each side of part H.
10. Stitch-weld all the long seams.

ATTACH THE HARDWARE

1. To attach the wheels to part G, drill a ⅛" hole ½" in from each end to accept a cotter pin. Your wheel preference will be determined by the surface you will be working on (concrete, dirt, or gravel). The example shown will allow for easy use of the cart on concrete floors.
2. Locate and weld two chain links to attach the cylinder safety chain.
3. Heat and bend both pieces of part H to a 3" radius. Place and weld the hooks on each side of the cart to safely hang cables.

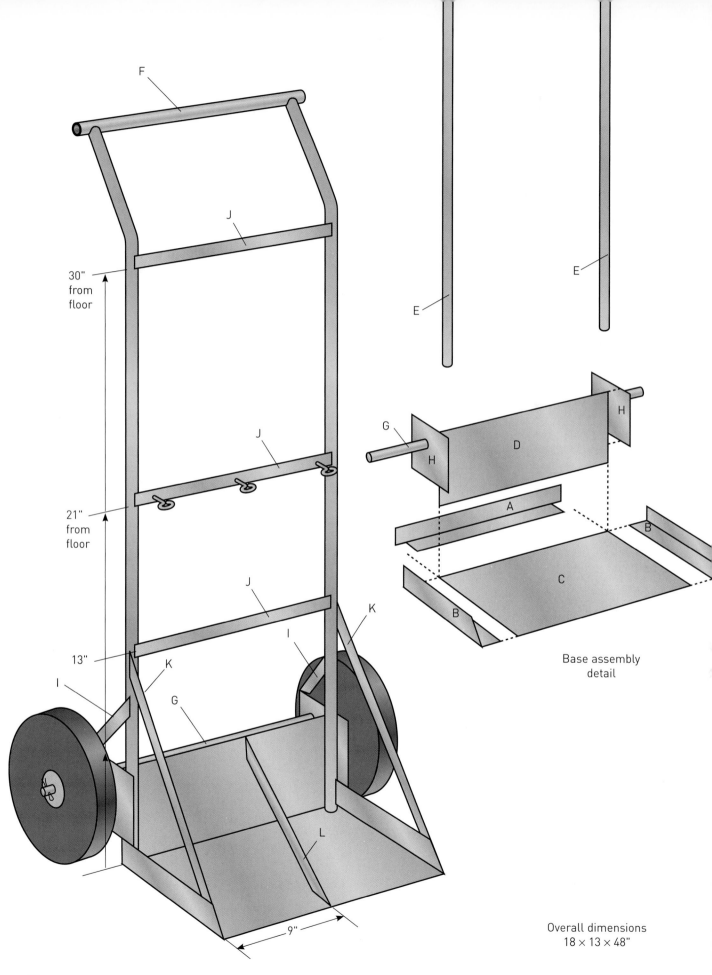

F

E

E

E

30"
from
floor

J

H

G

D

H

H

G

21"
from
floor

A

B

J

B

C

13"

J

K

I

K

I

B

Base assembly
detail

G

L

9"

Overall dimensions
18 × 13 × 48"

CYLINDER CART

You can purchase a cylinder cart for your oxyfuel rig—your dealer might even give you a discount—but it's fun and challenging to make your own. This cart is constructed with $1/8$" stock, which is oxyfuel weldable, but we used GMAW. The base platform and back support can be flame or plasma cut. If you have access to a heavy-duty metal brake, you can make the base platform and back support from one $16^{1}/_{4} \times 17^{3}/_{4}$" piece and bend it to 90° (allow $1/4$" for the radius of the bend).

MATERIALS

- $1/8 \times 1^{1}/_{2} \times 1^{1}/_{2}$" angle iron ($3^{1}/_{2}$')
- $1/8$" sheet metal (18×24" sheet)
- $1/8 \times 1$" round tube (12')
- $5/8$" round bar (2')
- $1/8 \times 1^{1}/_{2}$" flat bar (7')
- $1/8 \times 1^{1}/_{4}$" flat bar (4')
- 1" eyebolts with nuts (3)
- Chain
- Snap closures or threaded chain links (3)
- 8" wheels with hubs (2)
- Washers & cotter pins (2)

PART	NAME	DIMENSIONS	QUANTITY
A	Base back	$1/8 \times 1^{1}/_{2} \times 1^{1}/_{2}$" angle iron $\times$ 18"	1
B	Base sides	$1/8 \times 1^{1}/_{2} \times 1^{1}/_{2}$" angle iron $\times$ 10"	2
C	Base platform	$1/8$" sheet $\times$ 10 $\times$ $17^{3}/_{4}$"*	1
D	Back support	$1/8$" sheet $\times$ 6 $\times$ $17^{3}/_{4}$"*	1
E	Handle uprights	$1/8 \times 1$" round tube $\times$ 60"*	2
F	Handle	$1/8 \times 1$" round tube $\times$ 18"	1
G	Axle	$5/8$" round bar $\times$ 22"*	1
H	Axle brackets	$1/8$" sheet $\times$ 4 $\times$ $4^{1}/_{2}$"	2
I	Bracket supports	$1/8 \times 1^{1}/_{2}$" flat bar $\times$ 8"*	2
J	Crosspieces	$1/8 \times 1^{1}/_{2}$" flat bar $\times$ 17"*	3
K	Side supports	$1/8 \times 1^{1}/_{4}$" flat bar $\times$ 24"*	2
L	Base support	$1/8 \times 1^{1}/_{2}$" flat bar $\times$ 12"*	1

*Approximate dimensions, cut to fit

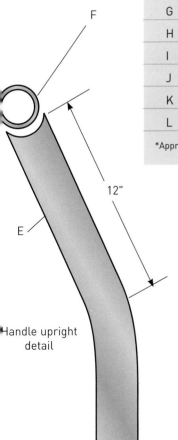

F

12"

E

Handle upright detail

HOW TO BUILD A CYLINDER CART

PREPARE THE CART BASE

1. Cut the base back (A) and base sides (B) to size. Rather than mitering the corners, cut a 1½" notch in each end of the base back (see diagram).
2. Mark and cut a triangular section off the front end of each base side to soften the edge.
3. Lay out the base back and sides to form a three-sided rectangle. Square the corners and clamp the assembly to your work surface.
4. Tack the joints at both ends of the base back. Turn the assembly over and weld the corner and butt joints.

INSTALL THE BASE PLATFORM & BACK SUPPORT

1. Measure the inside width of the base assembly and cut the base platform (C) and back support (D) to size.
2. Place the base platform into the base assembly and tack weld it into place along the base sides. (Do not weld along the base back, as that would prevent the back support from fitting flat against the base platform.)
3. Turn the base over and place two 1" welds along each side where the base platform meets the back and sides.
4. Set the base flat on the work surface, right side up, and place the back support upright against the inside of the base back (see photo, top left).
5. Clamp the back support to the base back. Place three 1" welds on the back side of the support where it meets the angle iron.
6. Place two 1" welds on the inside T-joint, making sure not to weld within two inches of the corner.

ATTACH THE AXLE BRACKETS

1. Cut the axle brackets (H) to size.
2. Place the base assembly on its side and align an axle bracket along the top edge of the base side, resting it on top of the back support and extending 3" off the back side.
3. Measure 4" up from the bottom of the base assembly and 1½" from the back edge to mark the axle location on the axle bracket. Drill, flame cut, or plasma cut a ⅝" hole at the mark.
4. Tack weld the axle bracket in place (see photo, bottom left). Insert the axle to check for proper alignment, then weld both sides of the T-joint between the bracket and the back support.
5. Weld the outside butt joint between the bracket and the base side.
6. Repeat steps 2 to 5 to weld an axle bracket to the other side of the base assembly.

PREPARE THE HANDLE UPRIGHTS

1. Cut the handle uprights (E) to length.
2. Make a 20° to 30° bend in one of the handle uprights 13" from one end using a heavy-duty conduit bender. Bend the other handle upright to match.
3. Cut both pieces so the curve is 12" from the top end and the overall upright height of each is 48". Grind the tops of both handle uprights so they fit around the handle tube.

After welding the base back and sides, cut the base platform and back support to fit, then weld them in place.

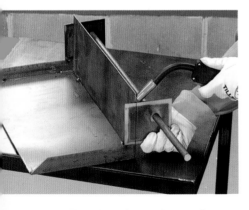

Tack the axle brackets to the base assembly. Insert the axle to check for proper alignment, then finish weld the bracket joints.

ASSEMBLE & INSTALL HANDLE UPRIGHTS

1. Place a handle upright against the inside corner of the base assembly. Turn the handle so the curve points straight back. Tack weld the handle in place. Repeat this process for the other handle upright.

2. Cut the handle (F) to size.

3. Place the cart on its back. Set the handle against the cutouts in the handle uprights keeping an equal amount of overhang on each end and tack weld in place.

4. Finish weld the handle uprights to the base assembly. Place welds between the handle uprights and top back support and between the handle uprights and axle brackets.

ATTACH THE CROSSPIECES

1. Measure and cut the crosspieces (J) so they are slightly recessed against the handle uprights.

2. Drill three $^3/_{16}$" holes—one in the center and one on each end of one crosspiece. Place this crosspiece 21" up from the base.

3. Set the remaining crosspieces at 13" and 30" from the base on the forward side of the handle uprights, and weld in place (see photo, top right).

ATTACH THE SUPPORTS

1. Turn the assembly on its side. Place a bracket support (I) in position across the handle upright and the axle bracket. Mark the angles on the support and cut it to size.

2. Weld the bracket support in place against the handle upright and axle bracket.

3. Set the side support (K) in position against the base side and the handle upright. Mark the angles, and cut to fit. Weld the support in place.

4. Turn the assembly on its opposite side and repeat steps 1 to 3.

5. Mark the angles and cut the base support (L) to size to fit between the base platform and the base back.

6. Center the base support from side to side on the base assembly. Weld both sides of each T-joint.

COMPLETE THE CART

1. Center the axle (G) between the axle brackets. Place the wheels on the axle to make sure the wheels barely make contact with the floor (see photo, bottom right).

2. Mark the axle for its final length and cut to size. Drill holes in each end of the axle to fit the cotter pins.

3. Center the axle between the axle brackets and weld in place.

4. Complete any unfinished welds.

5. Wire brush or sandblast the cart. Clean the cart, and paint as desired.

6. Attach the wheels to the axle. Insert cotter pins in the axle holes.

7. Affix eyebolts with nuts in the three holes in the middle crosspiece. Attach the chain to the eyebolts with snap closures or threaded chain links.

Lay the crosspieces across the handle uprights at the proper heights. Weld the pieces to the uprights.

Trial fit the wheels before welding the axle into place. The wheels should just barely make contact with the floor.

Top view

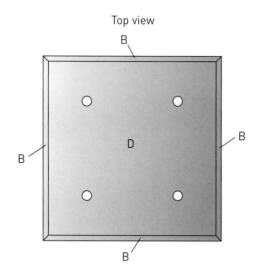

B

B

B

D

B

Overall dimensions
16 × 16 × 35"

MATERIALS

- ³⁄₁₆ × 1¹⁄₂ × 1¹⁄₂" angle iron (19')
- ³⁄₁₆" plate (8 × 8")

GRINDER STAND

This is a good project to make with angle iron from the discount bin at the local steel supply center. You can often buy short pieces for 10 cents a pound rather than the 30 to 50 cents you would normally pay. Take along your calipers or micrometer to measure thicknesses, as well as your measuring tape, cutting list, and tough gloves so you don't get cut digging through the scrap pile.

PART	NAME	DIMENSIONS	QUANTITY
A	Legs	³⁄₁₆ × 1¹⁄₂ × 1¹⁄₂" angle iron × 32"	4
B	Top	³⁄₁₆ × 1¹⁄₂ × 1¹⁄₂" angle iron × 8¹⁄₂"	4
C	Bottom	³⁄₁₆ × 1¹⁄₂ × 1¹⁄₂" angle iron × 16"	4
D	Platform	³⁄₁₆ to ¹⁄₂" plate 8 × 8"	1

HOW TO BUILD A
GRINDER STAND

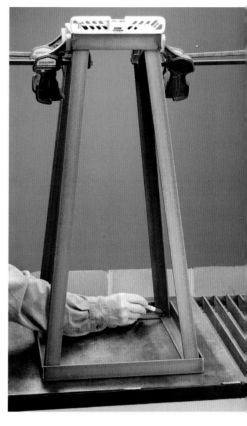

BUILD THE TOP & BOTTOM ASSEMBLIES

1. Cut the top and bottom pieces (B and C) to length. Miter the corners at 45° or cut 1½" notches into the pieces.

2. Assemble pairs of bottom pieces into right angles. Clamp the pieces to your work surface and tack weld at each corner.

3. Assemble the two right angles into a square. Check for square by measuring across each diagonal—if the measurements are equal, the assembly is square. Tack weld the corners.

4. Repeat steps 2 to 3 to assemble the top.

MARK & CUT THE LEGS

1. Cut the legs (A) to size.

2. Place the bottom assembly on your work surface and loosely clamp the legs into each corner.

3. Set the top assembly over the legs and clamp it in place. Make sure the top is level.

4. Mark each leg at the top and bottom where they cross the platforms (see photo, right).

5. Unclamp the legs and cut to size.

ASSEMBLE THE STAND

1. Place one leg on top of a corner of the bottom assembly and tack weld it in place. Tack weld the other three legs in the corners the same way.

2. Align a corner of the top assembly with the top of one leg and tack it in place. Tack weld the top assembly to the other three legs the same way.

3. Adjust the stand so the top is level. You may need to remove tack welds and grind down some angles.

4. When you are satisfied with the alignment, complete the welds. To minimize distortion, alternate between the sides, top, and bottom as you weld.

ATTACH THE PLATFORM PLATE

1. Cut the platform plate (D) to size. Drill four holes in the face to match your grinder mounting holes.

2. Set the plate over the top of the stand, aligning the edges with the outside of the top pieces. Weld the plate to the stand.

Assemble the grinder stand using clamps. Mark the top and bottom of each leg and cut to fit.

PORTABLE WELDING TABLE

Building a heavy-duty welding table helps you develop your welding skills, and it also provides a useful piece of equipment on which to build future projects. Because the table seen here is mounted on casters, it can be rolled away into a corner of the garage when not in use. The metal-slat tabletop can be covered with a removable sheet-metal welding surface that can be cleaned and reused as necessary. You can also build a plywood work surface for doing woodworking and metalworking that doesn't involve welding.

The welding table dimensions can be altered to meet the requirements of the welder. You could even build two tables of the same height to create an extended work surface. In short, this table is a very versatile tool, and the more you use it the more you'll appreciate it.

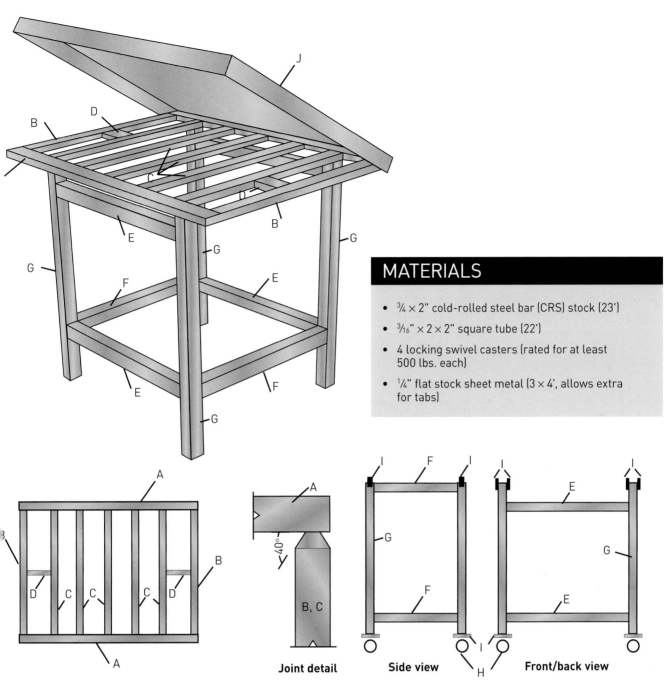

MATERIALS

- ¾ × 2" cold-rolled steel bar (CRS) stock (23')
- ³⁄₁₆" × 2 × 2" square tube (22')
- 4 locking swivel casters (rated for at least 500 lbs. each)
- ¼" flat stock sheet metal (3 × 4', allows extra for tabs)

Joint detail

Side view

Front/back view

PART	NAME	DIMENSIONS	QUANTITY
A	Top fronts	¾" × 2 CRS 40"	2
B	Top sides	¾" × 2 CRS 26"	2
C	Top crosspieces	¾" × 2 CRS 26"	5
D	Top short crosspieces	¾" × 2 CRS 4"	2
E	Front rails	³⁄₁₆" × 2" × 2" square tube 24"	4
F	Side rails	³⁄₁₆" × 2" × 2" square tube 22"	4
G	Legs	³⁄₁₆" × 2" × 2" square tube 34"	4
H	Casters		4
I	Tabs	2 × 2"	8
J	Tabletop	38 × 48 sheet metal	1

HOW TO BUILD A WELDING TABLE

Before welding, thoroughly clean all parts with denatured alcohol.

BUILD THE BASE SIDES

1. Measure and mark $3/16$" × 2 × 2 square tube for the four front rails (E) at 24" each, the four side rails (F) at 22" each, and the four legs (G) at 34" each.
2. Cut the pieces using a portable band saw or hardened-carbide-blade cut-off saw.
3. Align the side rails (F) flush in between two legs (see page 99): the uppermost side rail (F) is flush with the tops of the legs; the lower sides are 6" up from the bottoms of the legs. Use corner magnets to hold the lower sides square. To maintain your proper dimensions, use a 24" piece of scrap metal or 2 × 4 as a spacer between the two lower ends. Alternatively, tack weld a piece of metal in between the two pieces, and remove it later.
4. Tack the side rails in place. Repeat step 3 for the other side frame.
5. Align two front rails (E) between the side frame legs for the front: the uppermost front is 2" down from the top edge of the legs; the lower front is 6" up from the bottom edges of the legs. To maintain your proper dimensions, use a 22" piece of scrap metal or 2 × 4 as a spacer between the two lower ends. Alternatively, tack weld a piece of metal in between the two pieces, and remove it later. Tack a 6" scrap metal piece below the lowest part E so that the entire front/back frame can stand in place on its own as you tack the pieces to the legs.
6. Tack the front rails (E) to the legs in between both of the two side frames.

ADD THE TABS & CASTERS

1. Measure and mark $1/4$" flat stock for the tabs (I) at 2 × 2.
2. Check to make sure each tab is square, and then cut out eight tabs.
3. Align the tabs (I) no more than $11/16$" above the top of each leg (see page 97).
4. Hold the tabs in place with clamps, and then tack the tabs to the verticals. The top crosspieces (C) closest to the top ends (B) fit in between the tabs.

Measure and mark $3/16$" × 2 × 2 square tube for the four front rails (E) at 24" each, the four side rails (F) at 22" each, and the four legs (G) at 30" each.

5. Center casters over the ends of the legs.
6. Tack the casters, and then check for square. Adjust as necessary, and then weld. Use SMAW; 3/32 or 1/8" diameter E6013 electrode to weld on caster plates. Continuous welds are not necessary.
7. Weld all the way around each joint for the entire base frame tacked together thus far. *Note: Don't accidentally weld your temporary spacer bars that are used only to maintain dimension.*

BUILD THE TOP FRAME

1. Measure and mark two 3/4" cold-rolled steel bars to 40" for the top fronts (A). Cut them to length using a hardened-carbide-blade cut-off saw.
2. Also measure and mark 3/4" cold-rolled steel bars to 26" each for two top sides (B) and five top crosspieces (C).
3. Measure and mark 3/4" cold-rolled steel bars to 4" each for the two top short crosspieces (D).
4. Cut each of the pieces (A, B, C, D) for the tabletop frame following your cut lines.

5. Make the 40° end joints in each crosspiece (see page 97, Joint Detail) by grinding or flame-cutting the ends. The depth of each double bevel should be between 1/4" and 1/8".
6. Align the top sides (A) and top ends (B), and then check for square. Hold them in place with corner magnets.
7. Tack together all four sides. Use enough heat to get good fusion so when you move the frame the tacks won't break (metal is heavy).
8. Lay the top frame right-side up on a level table and start to align the crosspieces. Start in the center and evenly space each piece in between the two top sides (B). Make sure the crosspieces closest to the top sides are aligned so they fit in between the tabs that are welded onto the legs.
9. Align each crosspiece and hold it in place with corner magnets or clamps. Once everything is aligned properly, tack the cross pieces to the top fronts. *Note: 4.5" TYP indicates the "typical" spacing between the five middle crossbars is 4 1/2".*

(CONTINUED)

Cut the square tube using a metal-cutting bandsaw (shown) or cut-off saw with a hardened carbide blade.

Align the side rails in between legs. The side rail part is 6" up from the leg bottom. You will make two of these side frames and then add front rails in between the sides for the front and back of the table base.

10. Align the top short crosspieces in between the top sides (B) and crosspieces (C). Check for square, and then tack in place.

11. Check for final square of the entire tabletop frame, and make any final adjustments. Align the tabletop frame over the base to make sure the crosspieces fit in between the tabs on legs.

12. Complete all welds using sequence welding. Alternate from side to side, starting at the center crosspieces and moving to the outer crosspieces.

MAKE THE TABLETOPS

1. Measure and mark a 38 × 48" tabletop out of ¼" sheet metal.

2. Mark a line 4" in from the edge of each side. Also make the detailed marks for each corner bird's mouth. This will allow one edge of each corner to bend inside of the other edge of the corner for neat and strong edges.

3. Cut the tabletop sheet metal with portable shears (see photo, opposite, top).

4. Bend up the triangular bird's mouth tabs.

5. Use ⅛" steel as a bender. Lay the bender 4" in from the edges and then grasp the sheet metal and bender with clamps. Bend the sheet up at 90° (see photo, opposite, bottom left). The triangular bird's mouth tabs fold in to the inside for clean exterior corner edges.

6. Fit the tabletop over the top frame (see photo, opposite, bottom right).

FINISHING

1. Use a 4½" angle grinder with a 60-grit flapper disk to grind all welds until smooth.

2. Clean the metal, and then prime and paint the base, if desired. *Note: Do not paint the top.*

Measure and mark for the top short cross pieces. The tabletop frame should fit on top of the base with the cross pieces fitting in between the tabs on legs. The short cross pieces then fit in between the cross pieces and top sides (B) for extra strength.

Cut the sheet metal top to 38 × 48 using a portable metal shears.

Use a strip of ⅛"-thick steel as a bender. Clamp the bender to the tabletop sheet metal and bend up to 90°. Shown here: first make a bird's mouth cut on corners; bend those sides first so that the tabs are on the inside of the bend for a clean exterior edge.

Lower the finished tabletop onto the table frame.

SAW STAND

A saw table such as this is commonly used in workshops, home shops, and on job sites. Designed for use with a power miter saw or metal cut-off saw, it features outriggers that are adjustable, accommodating a variety of stock lengths. The outriggers slide out on both sides, opening to their maximum length of 24". If both sides need to be extended (for extra-long material), the outriggers can both be placed in the front and extend out of the same outrigger tubes. And this will be necessary because the back outriggers sit behind the saw—good for storage when the outriggers need to slide all the way in, flush with the stand sides (for moving and storage). You can add wood blocks on the ends of the outriggers to keep the object being cut level with the saw table.

MATERIALS

- 1¼ × 1¼" × 12 gauge steel (35')
- 8' of 1" × 1 × 12-gauge steel tubing
- 3/16" × 2" flat bar (2')
- ¼" round stock (1')
- Hex bolts and nuts ⁵⁄₁₆-18 UNC (4)
- ¾ to 1" thick Plywood (16 × 24")

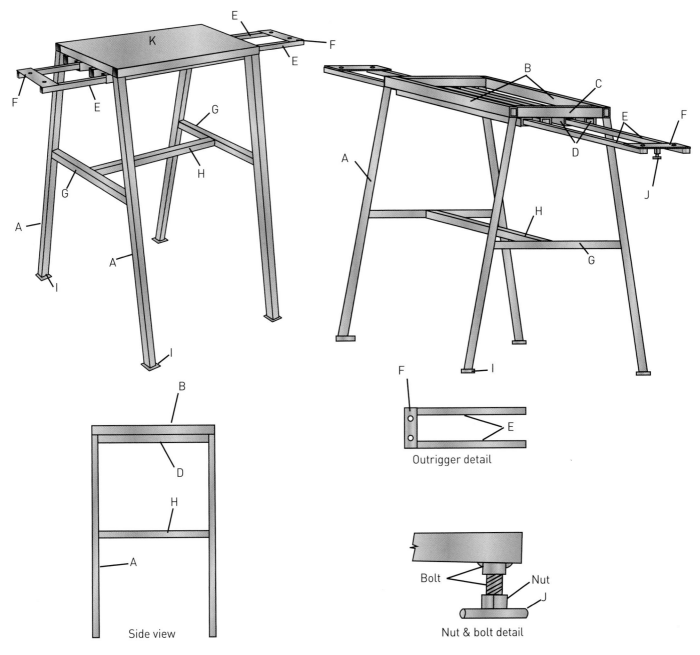

Outrigger detail

Side view

Bolt

Nut

Nut & bolt detail

PART	NAME	DIMENSIONS	QUANTITY
A	Legs	1¼ × 1¼", 12-gauge steel × 39"	4
B	Top long pieces	1¼ ×1¼", 12-gauge steel × 24"	2
C	Top short pieces	1¼ × 1¼", 12-gauge steel × 13½"	2
D	Outrigger tubes	1¼ × 1¼", 12-gauge steel × 24"	4
E	Outriggers	1 × 1, 12-gauge steel × 24"	4
F	Outrigger end	3/16 × 2" flat bar × 7¾"	2
G	Leg connectors	1¼ × 1¼, 12-gauge steel × 20" (at top edge)	2
H	Leg connector crosspiece	1¼ × 1¼, 12-gauge steel × 21½"	1
I	Leg feet	3/16" × 2" flat bar × 2 × 4 (approx.)	4
J	Outrigger handle	3 × ¼" round stock	4
K	Tabletop	16 × 24" plywood	1

HOW TO BUILD A SAW STAND

Tack together the top long (B) and short (C) parts.
Shown here: MIG welder.

Tack the ends of the legs (A) cut to 10° onto the top frame.

BUILDING THE TOP FRAME

1. Cut the top long pieces (B) to 24".
2. Cut the top short pieces (C) to 13½".
3. Align the top long pieces (B) and top short pieces (C) at 90°. Use a corner magnet to hold the pieces together squarely.
4. Check for square again. Use a magnetic clamp or simply hold the parts in place as you tack them together. Tack all of the long and short tops together to create the top frame.
5. Check for square again, and then weld the four pieces together.

Note: Recommended methods for all welds: SMAW 3/32" E 6013 electrode; GMAW short circuit transfer; GTAW; ER 70S-6 filler metal, or MIG (as shown here).

BUILDING THE SIDES

1. Cut the legs (A) out at 39" with parallel 10° ends (see page 103).
2. Align the 10° cut end of a leg (A) 90° to a top long piece (B).
3. Check the pieces for square, and then hold the legs in place 90° against the top long piece. Tack the leg in place.
4. Repeat steps 2 and 3 for the other leg on the same side.
5. Repeat steps 2 through 4 for the other side.
6. Cut the leg connectors (G) to length at 20" with a 10° angle on each end (from the top edge of the the angle to the top edge of the opposite angle it should measure 20").

Tack the horizontal leg connectors (G) and then center cross piece (H) that spans between the two leg connectors.

Flip the entire stand over and weld the leg feet (I) to the bottom side of the legs (A). If desired, casters could be added to the bottom for increased mobility. Be sure to pick casters that can lock.

7. Hold a leg connector (G) in between two legs. Check for proper alignment, and then tack in place.
8. Check for square, and then weld the connectors in place.
9. Repeat the above steps for the leg connector (G) on the other side.
10. Center the leg connector crosspiece (H) in between each leg connector (G). Align at 90°, and then use a magnetic clamp to hold it in place. Check for square, and then tack the crosspiece to the two leg connectors.
11. Check for square, and then weld in place.
12. Cut all four feet (I), each at 2".
13. Flip the table upside down on the floor, and then align the feet under the table legs. Hold them in place with pliers, and then tack them in place.
14. Flip the entire table right-side up to check for level. Adjust as necessary, breaking the feet off and cutting legs for level, if necessary. Shims could also be added.
15. Weld the feet to the legs.

BUILDING THE OUTRIGGERS

Note: 12-gauge square tubing will always slide, one tube inside the other, when they are ¼" difference in size. More precisely, in order for the outriggers to slide into the outrigger tubes, 12-gauge tubing must be used. 12-gauge tubing has a wall thickness of .110" to .115", so tubes will slide into each other in ¼" increments. For a tighter fit, 11-gauge seamless tubing can also be used, but seamless tubing is more expensive.

1. Cut outrigger tubes (D) from 1¼ × 1¼", 12-gauge steel to length at 24".
2. Cut the outriggers (E) from 1 × 1, 12-gauge steel to length at 24". Drill a single ⅜" hole on one end of each outrigger. This hole will eventually allow the screw/bolt on the outrigger tube to screw up into the extended outrigger, securely holding it in place when fully extended.
3. Cut the outrigger ends (F) to size at 7¾". Drill two ⅜" holes in them, 2" in from each end.
4. Align an outrigger end (F) 90° to outriggers 1 and 3 (E) at the opposite end of the ⅜" hole you drilled in step 2.
5. Align the other outrigger end 90° to outriggers 2 and 4 (also at the opposite end of the ⅜" hole you drilled in step 2).
6. Use a carpenter's square to ensure proper alignment, and then clamp in place. Tack an outrigger end to outriggers 1 and 3, then repeat for outriggers 2 and 4. Check for square, and then weld the outrigger ends in place. (CONTINUED)

Drill ⅜" holes in the outrigger ends (F). Use a scrap piece of wood under the flat bar, and clamp the bar and wood to a table.

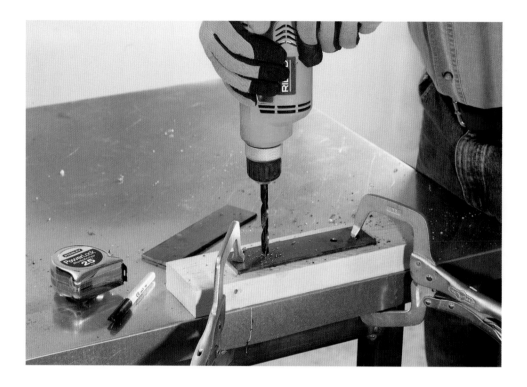

7. Drill a ⅜" hole on one end of each outrigger tube (D) underside.
8. Hold the outrigger handle (J) in place over bolt end, and then weld in place.

FASTENING OUTRIGGERS TO FRAME

1. Center a nut over each of the outrigger tube (D) holes from step 7 (see Nut & Bolt Detail on page 103). Use pliers to hold the nut in place, and then weld. Make sure to alternate: nuts should be on the same end of 1 and 3 outrigger end (F); and 2 and 4 outrigger end (F). The nut is welded onto the tubing after drilling a hole in it. The bolt that screws in is used as a locking mechanism.
2. Clamp outrigger tubes (D) to the bottom side of the top short pieces (C) with 2" spacing. Check alignment with a ruler.
3. Insert the outriggers (E) into the outrigger tubes (D) so that 1 and 3 extend to one side and 2 and 4 extend to the opposite side. Check for ease of sliding.
4. Once the slides are positioned so that they can be pulled out easily, remove them and tack the outrigger tubes onto the top short pieces (C).
5. Double check to see that outriggers slide in and out with ease. If they do, completely weld the outrigger tubes in place. Weld only two sides of the connections, where tubes connect together. No need to weld all around it.

ADDING THE TOPS

1. Cut a 16 × 24" tabletop (K) out of ¾"- to 1"-thick plywood.
2. Predrill ⁵⁄₁₆" holes for the plywood tabletop (K) into the frame (B, C). Screw or bolt the tabletop in place.

OPTION: Cut 2 × 2" wood blocks for the outrigger tops (M). These sit directly above the holes you previously drilled in the outrigger ends (F). Predrill holes in the wood blocks (see photo, left), and then screw the blocks onto the outrigger ends (F). These blocks keep your cutting object level as it spans out on the outriggers.

FINISHING

1. Cap all ends of exposed tubing.
2. Prime and paint the saw table.
3. Prime and paint the table frame.
4. Completely remove outriggers and prime them. Once dry, they should be painted or powder-coated for extra durability. For optimal rust protection, first roughen the surfaces by sanding and apply a rust-resistant metal primer before painting.

Tack weld the outrigger ends (F) to the outriggers (E). There are two complete pullout outriggers for each side of the stand.

Tack weld the outrigger tubes (D) to the top frame underside.

Fasten the plywood top to the frame.

HOME DÉCOR

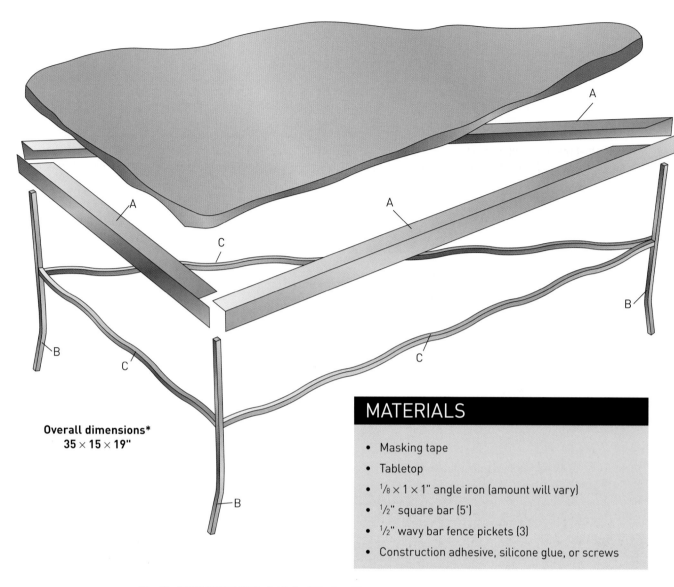

Overall dimensions*
35 × 15 × 19"

COFFEE TABLE BASE

A nice slab of stone or a cross section of a tree makes an appealing coffee table, but how do you support it simply and easily? A table base welded from 1" angle iron and $\frac{1}{2}$" square bar can support the weight of stone or wood without being cumbersome or bulky. This design is for a triangular base to suit the piece of marble being used. Creating a layout with tape allows you to experiment with different sizes and locations for your triangle. If you are using fence pickets, as we have here, make sure the length of the longest side of the triangle does not exceed the length of the picket. Pickets are generally 35" to 39" long.

PART	NAME	DIMENSIONS	QUANTITY
A	Top supports	$\frac{1}{8} \times 1 \times 1$" angle iron*	3
B	Legs	$\frac{1}{2}$" square bar × 20"	3
C	Wavy crosspieces	$\frac{1}{2}$" wavy bar fence pickets*	3

*Side pieces and top supports must be cut to fit the particular tabletop.

HOW TO BUILD A COFFEE TABLE BASE

CREATE THE TRIANGLE

1. Turn the tabletop upside down and use masking tape to lay out a triangle. The triangle should be at least 3" in from the edges of the tabletop material, but not so far in that the table will tip easily.
2. Measure each side of the triangle. Cut the top support pieces (A) to match these three measurements. Miter both ends of the longest piece.
3. Place the longest support piece on top of the second-longest support. Set both pieces on the tape layout on the tabletop. Mark the angle and notch on the second support where it intersects the first support (see photo) to allow these pieces to butt together. Cut out the notch and angle on the support.
4. Place the two supports back together on the triangle layout. Set the third support on the layout under the first two support pieces. Mark the notches and angles at the intersections. Cut the third support piece. (You may need to make an additional angle cut on the second long piece, depending on the triangle.)
5. On a work surface, arrange the support pieces to form the triangle. Weld the outside corners. Turn the assembly over and weld the top butt joints.
6. Use an angle grinder to smooth the top joints so the tabletop rests on a flat surface.

INSTALL THE LEGS

1. Cut the legs (B) to size.
2. Mark each leg 4" from one end. Place a leg in a bench vise, lining up the mark with the edge of the vise jaws and bend the leg end 15°. Repeat this process for the other two legs.
3. Turn the top assembly upside down on your work surface. Place a leg in a corner with the bend of the leg pointing outward in the same line as the point of the triangle.
4. Check to make sure the leg is perpendicular to the top assembly, then tack weld in place. Tack weld the other two legs in the remaining corners the same way.

ATTACH THE SIDE PIECES

1. Measure the distance between the legs. Cut one wavy crosspiece (C) to fit each of the three sides. Bevel the ends to fit the angled legs.
2. Clamp a crosspiece against two legs 7" from the top assembly. Check for square, and tack weld. Install the remaining crosspieces the same way.
3. Turn the assembly right side up, and check for square and level. Make adjustments if necessary. Complete all the welds.

COMPLETE THE TABLE

1. Grind down the welds, if desired. (You may also want to grind down the feet of the table so they make flat contact with the floor.) Clean or wire brush the assembly. Apply your choice of finish.
2. Drill holes in the angle iron and attach a wooden tabletop with screws, or use silicone or construction adhesive to attach a stone top.

Place the mitered long support piece on top of the second-longest support piece. Mark the angle and notch needed to fit the pieces together.

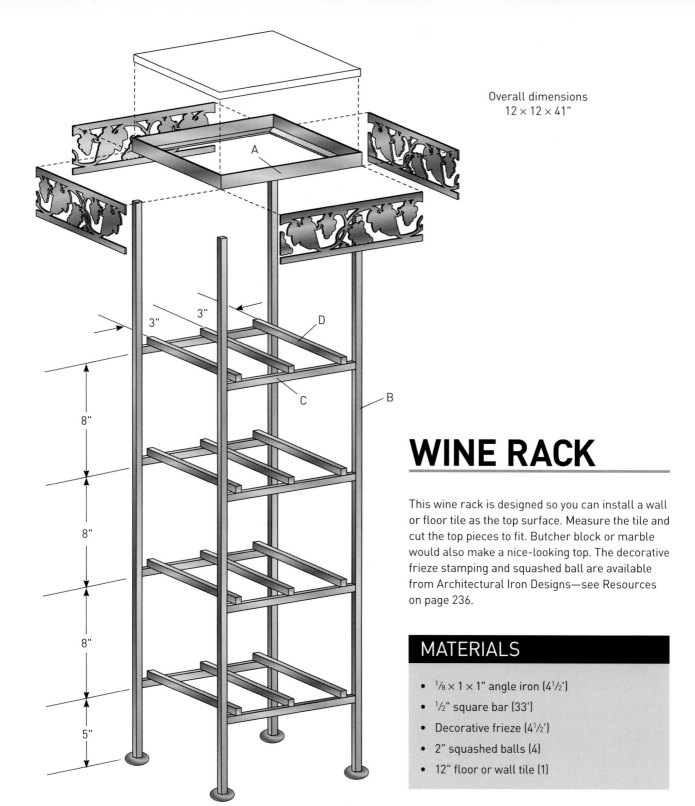

Overall dimensions
12 × 12 × 41"

WINE RACK

This wine rack is designed so you can install a wall or floor tile as the top surface. Measure the tile and cut the top pieces to fit. Butcher block or marble would also make a nice-looking top. The decorative frieze stamping and squashed ball are available from Architectural Iron Designs—see Resources on page 236.

MATERIALS

- ¹⁄₈ × 1 × 1" angle iron (4¹⁄₂')
- ¹⁄₂" square bar (33')
- Decorative frieze (4¹⁄₂')
- 2" squashed balls (4)
- 12" floor or wall tile (1)

PART	NAME	DIMENSIONS	QUANTITY
A	Top	¹⁄₈ × 1 × 1" angle iron × 12"*	4
B	Legs	¹⁄₂" square bar × 40"	4
C	Rack supports	¹⁄₂" square bar × 11" *	8
D	Bottle holders	¹⁄₂" square bar × 12" *	12

*Cut to fit selected tile top

HOW TO MAKE A WINE RACK

ASSEMBLE THE TOP

1. Cut the top pieces (A) to size to fit your top surface, mitering the corners at 45°.
2. Clamp two top pieces together with a corner clamp. Use a carpenter's square to check the pieces for square, and tack weld.
3. Repeat step 2 to assemble the other two top pieces.
4. Clamp the two L-shaped top pieces together, and check for square by measuring across both diagonals. If the measurements are equal, the assembly is square. If it is not square, adjust until both measurements are equal.
5. Tack weld the two corners. Unclamp the pieces and recheck for square. Finish welding the corner joints.

ASSEMBLE THE STAND

1. Cut the legs (B), rack supports (C), and bottle holders (D) to size.
2. Mark the legs at 5", 13", 21", and 29" from one end. Mark the rack supports at 2½", 5½", and 8½" from one end.
3. Position four rack supports between two of the legs at the marks. Check the pieces for square and tack weld in place.
4. Repeat step 3 to tack weld the remaining four rack supports to the other two legs.
5. Stand the leg assemblies upright and clamp a bottle holder centered on the 5½" mark on the bottom rack support. Clamp the other end of the bottle holder to the other leg assembly. Check for square and tack weld the bottle holder in place (see photo).
6. Repeat step 5 to attach a bottle holder at the 5½" marks on the other three rack supports (see photo).
7. Make sure the leg assemblies and bottle holders are square, then add the remaining bottle holders at the 2½" and 8½" marks. Finish all welds.

ATTACH THE TOP, LEGS & FRIEZE

1. Place the top onto the legs and weld in place.
2. Center a squashed ball at the bottom end of each leg and weld in place.
3. Bend or cut the decorative frieze to fit around the top of the legs and top. Braze, braze weld, or tack weld the frieze into place at the top of the legs and slightly below the top.
4. Grind down the welds as needed. Wire brush or sandblast the wine rack, then apply the finish of your choice.
5. Place the tile into the top.

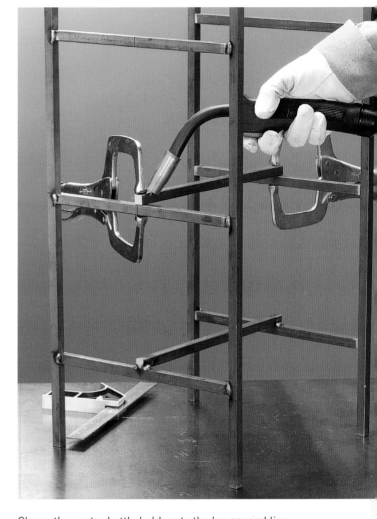

Clamp the center bottle holders to the leg assemblies and tack weld in place.

SWIVEL MIRROR

Reflect your great taste with this sturdy modern mirror. Its classic design creates a mirror built to last and will make a stunning first impression when displayed in your front entryway for years to come. The clean lines and black paint give this mirror a modern feel, but this can easily be altered by the finish treatment you choose. The adjustable mirror allows even the most petite guests to swivel it to the perfect height.

As described here, the inside frame is welded to the outside frame from the back using numerous small welds. The joints between the inside frame pieces are welded from the front, and those welds are ground down. The final look is a frame within a frame, rather than a completely smooth single frame. If you prefer a solid frame, you may continuously weld the inside frame to the outside frame from the front and then grind the welds smooth.

MATERIALS

- 16-gauge 1 × 1" square tube (22')
- $^1/_8$ × $^1/_2$" flat bar (11')
- $^1/_2$ × $^1/_2$" square bar (7")
- $1^1/_2$" × $^5/_{16}$" hex bolts (2)
- $^1/_4$"–20 hex nuts (6)
- Levelers (4, Rockler #32498)
- $^1/_2$ × 1" flat bar (6")
- Mirror (45$^7/_8$" × 17$^7/_8$")
- $^1/_4$" fiberboard (46" × 18")

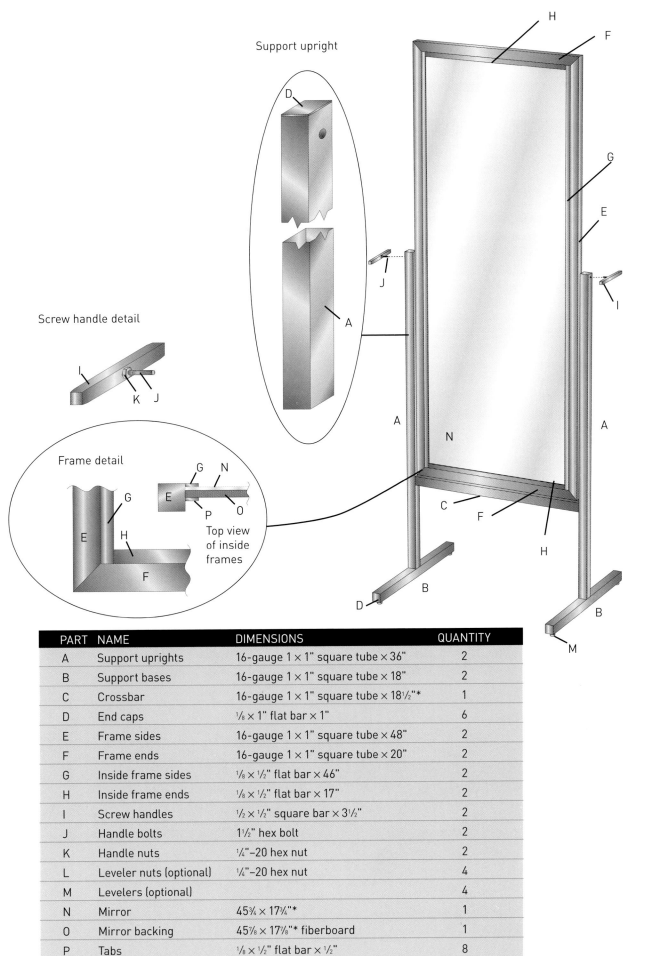

Support upright

Screw handle detail

Frame detail

Top view
of inside
frames

PART	NAME	DIMENSIONS	QUANTITY
A	Support uprights	16-gauge 1 × 1" square tube × 36"	2
B	Support bases	16-gauge 1 × 1" square tube × 18"	2
C	Crossbar	16-gauge 1 × 1" square tube × 18½"*	1
D	End caps	⅛ × 1" flat bar × 1"	6
E	Frame sides	16-gauge 1 × 1" square tube × 48"	2
F	Frame ends	16-gauge 1 × 1" square tube × 20"	2
G	Inside frame sides	⅛ × ½" flat bar × 46"	2
H	Inside frame ends	⅛ × ½" flat bar × 17"	2
I	Screw handles	½ × ½" square bar × 3½"	2
J	Handle bolts	1½" hex bolt	2
K	Handle nuts	¼"–20 hex nut	2
L	Leveler nuts (optional)	¼"–20 hex nut	4
M	Levelers (optional)		4
N	Mirror	45¾ × 17¾"*	1
O	Mirror backing	45⅞ × 17⅞"* fiberboard	1
P	Tabs	⅛ × ½" flat bar × ½"	8

* Approximate measurement, cut to fit.

HOW TO BUILD A SWIVEL MIRROR

Before welding, thoroughly clean all parts with denatured alcohol.

ASSEMBLE THE SUPPORTS

1. Cut the support uprights (A) and support base pieces (B) to length.
2. If using levelers, drill two $11\frac{7}{64}$" holes in one side of each loose piece, 2" from the ends. Drill through one wall only. Weld $\frac{1}{4}$" hex nuts over the holes.
3. Center an upright on a base at 90° and check for square. Weld into place, and repeat with the second upright and base.
4. At 2" from the top outside of each upright, drill a $1\frac{1}{32}$" hole. Drill the hole through both walls of the tube.
5. Cut end caps (D) to size. Weld caps to the support upright and base ends (see photo, opposite top). The two supports are in the shape of a T, thus creating the need for six end caps.
6. Grind all welds smooth.

ASSEMBLE THE FRAME

1. Cut the frame sides (E) and frame ends (F) to size. Miter the ends at 45°, or create a joint as shown in the Frame Detail (on page 115).
2. Drill a $1\frac{1}{32}$" hole at the outside midpoint of each side through one wall for the mirror pivot bolts.
3. Align a frame side and a frame end to make a 90° angle. Use a carpenter's square to check for square, and clamp in place. Tack weld the pieces together to form an L. Repeat with the other frame side and frame end.
4. Align the two Ls to form a rectangle. Use a carpenter's square to check for square. Clamp into place. Measure the diagonals of the rectangle to check for square. The diagonal measures should be equal; if not, adjust until they are. Tack weld together and recheck for square.
5. Complete all the frame welds. Grind down each weld until flush.
6. Grind the zinc coating off the handle nuts (K). Center the nuts over the holes and weld into place. Protect threads with anti-spatter gel (see photo, opposite bottom left).

ASSEMBLE THE INSIDE FRAME

1. Before cutting the inside frame sides (G) to length, measure the inside length of the assembled frame. Cut the inside frame sides to fit.
2. Place the assembled frame on a flat surface. Place the two inside frame sides in place against the two frame sides and tack weld into place.
3. Measure between the inside frame sides, and cut the inside frame ends (H) to length.
4. Place the inside frame ends in place against the frame ends and tack weld into place.
5. Turn the frame unit over. The inside frame should be flush with the frame. If not, remove pieces and retack, as necessary.
6. Weld the inside frame to the frame from the back with small welds every 3" to 4" (see photo opposite bottom right). Weld the joints between the inside frame members from the front.
7. Grind down all the welds on the front so they are flush.

(CONTINUED)

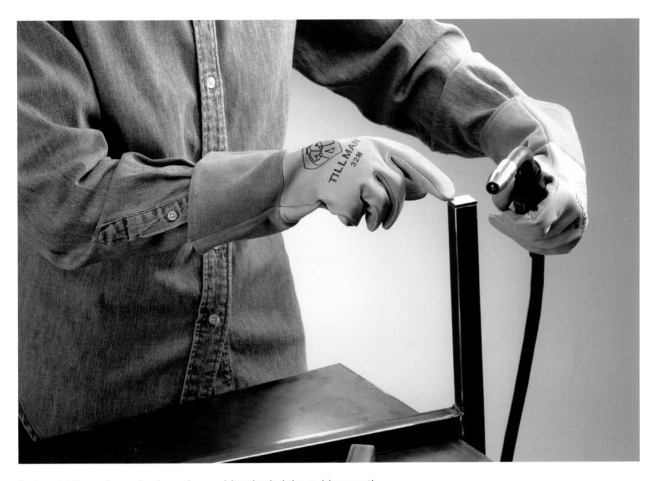

Tack weld the end caps in place, then weld and grind the welds smooth.

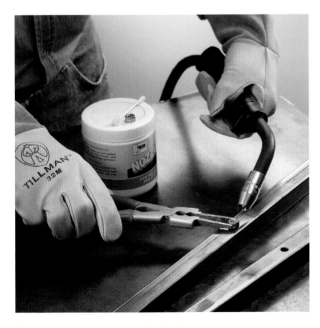

Weld the nuts over the holes in the frame sides before assembling the frame. Use anti-spatter gel to protect the nut threads.

From the back, weld the inside frame to the frame. A small weld every 3" to 4" is sufficient.

Weld the bolts to the beveled screw handles. Use masking tape or anti-spatter gel to protect the threads.

MAKE THE HANDLES & APPLY FINISH

1. Cut the screw handles (I) to length. Bevel the ends.
2. Grind the zinc coating from the head of the handle bolt (J). Coat the bolt threads with anti-spatter gel or cover with masking tape. Weld the bolts to the center of the bolt handles (see photo, above).
3. Assemble the frame and supports. With the bolts tightened enough to prevent the frame from moving, measure the distance between the support uprights making sure the uprights are parallel during measurement. Cut the support crossbar (C) to this length.
4. Measure the distance between the frame and the uprights (it should be about ¼"). Mark the uprights below the frame at that same distance. The top of the crossbar will align with these marks so the spacing between the frame and uprights and crossbar is uniform.
5. Disassemble the frame and supports. Align the top of the crossbar with the marks on the uprights. (This is easiest with support bases hanging over the edge of the work surface.) Tack weld the crossbar into place (see photo, opposite top). Use a carpenter's square to check that all angles between the crossbar and support uprights are 90°. Complete the welds and grind them smooth.
6. Grind all welds smooth, and grind off any spatter. Wipe the parts down with denatured alcohol to remove grinding dust. Prime and paint. For a natural metal look, use a clear coat. Protect bolt and nut threads with tape during painting to prevent fouling. Don't forget to paint the handles.

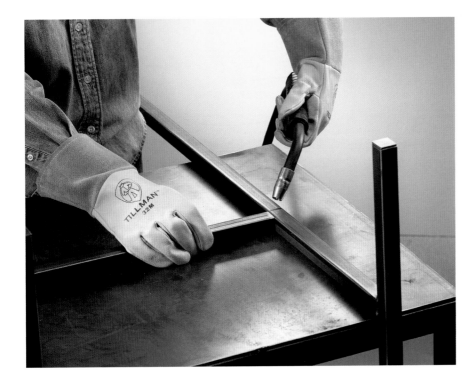

Assemble the frame and uprights, and mark the location for the crossbar. Remove the frame and weld the crossbar in place.

INSTALL THE MIRROR

1. Measure the inside of the completed frame to determine the exact mirror size to order. The mirror (N) should be ¼" shorter and ¼" narrower than the frame.
2. Cut the mirror backing (O) to size.
3. Apply small dots of silicone adhesive every 12" around the inside frame. Place the mirror into the frame.
4. Apply small dots of silicone adhesive to the back of the mirror. Place the backing against the mirror. Place the mirror assembly face down on a non-scratching surface.
5. Cut the tabs (P) to size, and grind one end to a semi-circle.
6. Space the tabs at 6", 23", and 40" along the frame sides. Sand a small amount of paint off the frame at these points. Center a tab along the frame top and bottom.
7. Firmly push the tabs against the mirror backing and frame sides. Weld into place (see photo, right).
8. Paint the tabs and the smooth side of the mirror backing to match the frame.
9. Install the levelers, if desired, and assemble the mirror. Position the mirror within the uprights, and thread the handle bolts through the support uprights and into the frame sides.

Install the mirror and back in the frame. Sand the finish from the frame and weld the tabs.

COAT RACK

This chic coat rack has a modern artistic flair due to the unique circular patterns. But the basic design has unlimited potential. You can customize this rack for any part of your life—the front entryway, the garage, the gardening shed, and the cabin may all have different styles. Simply replace the spirals with a punched tin pattern, decorative frieze, or dragonfly, moose, or bear stampings. The rack is sized to mount on studs spaced 16 inches apart. If you like the spirals but want more contrast, use copper or brass 14-gauge wire and leave the entire piece unpainted—but be sure to braze the spirals into place, instead of welding them.

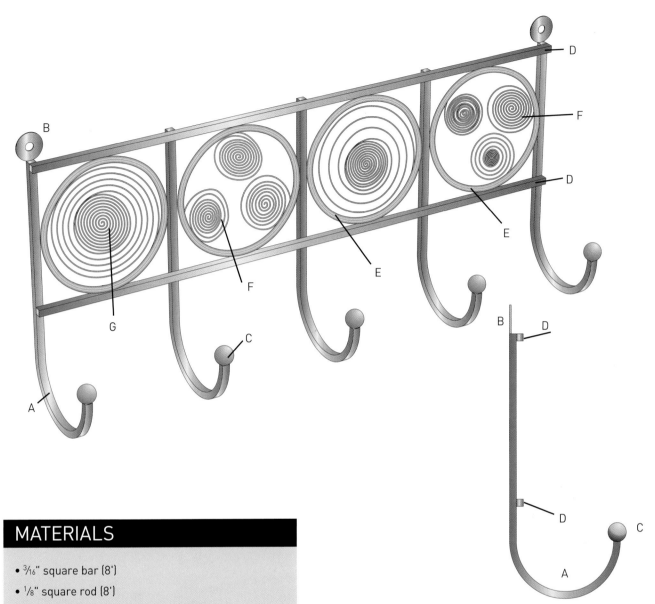

MATERIALS

- $\frac{3}{16}$" square bar (8')
- $\frac{1}{8}$" square rod (8')
- $\frac{1}{2}$" balls (5, decorative iron #JG8851)
- $\frac{3}{4}$" washers (2)
- 18-gauge annealed stove-pipe wire (25')

PART	NAME	DIMENSIONS	QUANTITY
A	Hooks	$\frac{3}{16}$" square bar × 11½"	5
B	Mounting washers	$\frac{3}{4}$"	2
C	Balls	$\frac{1}{2}$"	5
D	Crossbars	$\frac{3}{16}$" square bar × 16"	2
E	Circles	$\frac{1}{8}$" square rod × 24"	4
F	Mini-spirals	18-gauge wire	6
G	Spirals	18-gauge wire	2

HOW TO BUILD A COAT RACK

Before welding, thoroughly clean all parts with denatured alcohol.

MAKE THE HOOKS & FRAME

1. Cut the hooks (A) to length.
2. Clamp one end of each hook to a 2"-diameter pipe with a vise-style pliers. Bend the bar a half turn around the pipe to create each hook.
3. Weld each of the two mounting washers (B) to the top of two separate hooks (see photo, below). Weld a ball (C) to the end of each hook.
4. Cut the crossbars (D) to length. Mark the crossbars at 4", 8", and 12".

5. Align the two hooks that have the mounting washers flush with the crossbar ends. Tack weld in place. Place the second crossbar 4" from the top crossbar and tack weld in place. Check that the angle between the crossbars and hooks is 90°, and finish the welds.
6. Center a hook under the two crossbars at the 12" mark lining the top of the hook flush with the top crossbar. Check that the angle between the hook and crossbars is 90°, and weld into place.
7. Weld the remaining two hooks to the crossbars at 4" and 8" (see photo, opposite left).

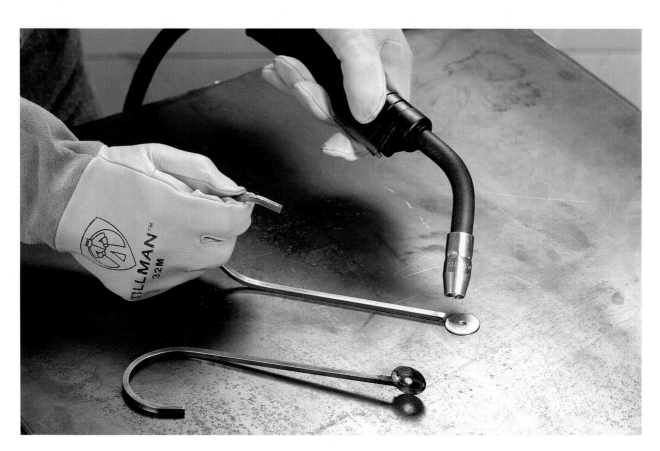

Weld the mounting washers to two of the hooks.

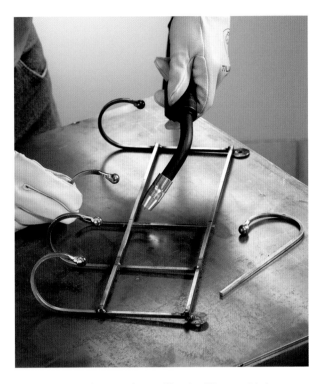

Weld the crossbars to the end hooks. Then weld the remaining hooks centered at 4", 8", and 12".

Begin a spiral by wrapping the wire around the tip of a needlenose pliers. Then grip the circle in the pliers and bend the wire until the desired size spiral is achieved.

MAKE THE CIRCLES & SPIRALS

1. Cut the circles (E) to length.
2. Clamp one end of a circle blank to a 4" pipe using a vise-style pliers. Bend the rod around the pipe one and a half turns to form a circle. Repeat to form the remaining circles.
3. Create the spirals (F) and mini-spirals (G) using a needlenose pliers. Wrap the wire around the end of the pliers to create a circle. Remove the circle from the pliers and grip it in the pliers. Wrap the wire around the circle until the desired size spiral has been formed (see photo, above right).

ASSEMBLE THE DECORATIVE FEATURES

1. Place a circle in the space between the first two hooks starting either on the far left or far right end. Squeeze or expand the circle to fit into the space and cut off any extra length. Weld the circle ends together and weld the circle to the two hooks.
2. Repeat step 2 with the remaining circles.
3. Arrange three mini-spirals inside the second and fourth circles. Weld them to each other and to the circle at contact points.
4. Place the large spirals between the remaining hooks and weld in place.
5. Wire brush or sandblast the coat hanger. Apply the desired finish.

ROOM DIVIDER

Folding screens are versatile interior design pieces. Used as an accent behind furniture, as a room divider, or as a way to hide an unsightly outlet or wall blemish, this piece holds its own as a piece of art but is also practical furniture. In large spaces it provides a focus while also creating a seamless flow throughout the room. But don't limit this folding screen to indoor spaces. The all-metal version presented here could easily be incorporated into a garden as a trellis or fence panel.

This project includes options. You can make a gathered fabric panel or a decorative metal panel. Or you could create both together. As constructed, this screen is moderately sized. Measure the area where the screen will reside and change the proportions accordingly.

Any number of decorative options can be used for the screen panels. Stained glass and pierced or expanded sheet metal are options not covered here, but they would be fabulous additions. If you spend some time with the catalogs of the decorative metal suppliers (see Resources, page 236) you will discover many leaf and flower stampings that could be used to create your own personalized pattern. Grapevines, oak leaves, and sunflowers are available in many different sizes; animal figures, sporting figures, and numbers and letters are available as well.

NOTE: If you don't care to cut out the leaves as listed in this project, use a small dapped leaf by Architectural Iron Designs. Try to find a leaf in a similar pattern.

MATERIALS

- 16-gauge ¾" round tube (49')
- 16-gauge 1" round tube (2')
- ¼" round rod (6', optional)
- Fabric (5 yards)
- 16-gauge ½" round tube (28')
- ⅛" round rod (14')
- 22-gauge sheet metal (12 × 24")

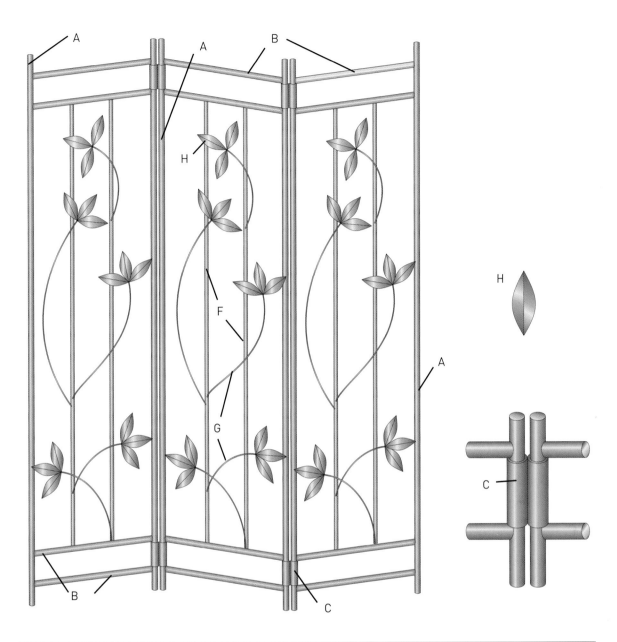

PART	NAME	DIMENSIONS	QUANTITY
A	Legs	16-gauge ¾" round tube × 68"	6
B	Crossbars	16-gauge ¾" round tube × 15"	12
C	Hinges	⅛ × 1" round tube × 3"	8
D	Fabric hanging rods (optional)	¼" round rod × 15½"	4
E	Fabric (optional)	60 × 30"	2
F	Verticals	16-gauge ½" round tube × 54¾"	6
G	Stems	⅛" round rod*	15
H	Leaves	22-gauge sheet metal (4 × 1½"*)	45

* Approximate measurement, cut to fit.

HOW TO BUILD A ROOM DIVIDER

Before welding, thoroughly clean all parts with denatured alcohol.

MAKE THE FRAMES

1. Cut the legs and crossbars (A and B) to length.
2. Mark each leg at 6" in from each end.
3. Using a bench or angle grinder, grind the end of each crossbar to shape it to fit around the leg. Make sure the curves are oriented in the same direction at both ends (see photo, below).
4. Place two legs on a large work surface. Align the bottom of one crossbar and the top of another with the 6" marks. Tack weld the crossbars in place. Check the frame for square and weld in place. Grind the welds smooth.
5. Repeat step 4 for the remaining two panels.

ADD THE HINGES & COMPLETE THE FRAMES

1. Cut the hinges (C) to length.
2. Slide the hinges over all four leg ends of one panel and the top and bottom leg ends of the one side of the other two panels.
3. Align the second set of crossbars above and below the hinges and tack weld in place. Check for square, and weld. Make certain you are not welding the hinges to the legs as you weld the crossbars.

OPTION 1: CREATE A FABRIC PANEL

The fabric panel is gathered. If you prefer a smooth look, decrease the fabric width to 15½".

MAKE THE HANGING RODS

1. Cut the hanging rods (D) to length.
2. Mark points on the inside of each panel leg, ½" down from the inside top and bottom crossbars.
3. At the marked points, drill a ¹⁷⁄₆₄" hole through the leg wall.
4. Insert one end of a hanging rod into a hole until it contacts the opposite wall. It should then be possible to insert the other end of the hanging rod into the hole on the opposite leg. If it is too long, grind it down until it fits.
5. Grind down any rough welds and grind off spatter. Wipe down with denatured alcohol to remove grinding dust. Finish as desired.

Grind a semicircle in the crossbar ends to create a saddle to fit around the legs.

MAKE THE FABRIC PANEL

1. Make a ¼" flat-felled hem on the long sides of the fabric.
2. Drape the fabric over the top hanging rod so that equal amounts are extending past the top and bottom hanging rods.
3. Pin the fabric together around the rods.
4. Remove the rods and sew the rod pocket, turning the raw edge under.
5. Slide the fabric over the rods and reinstall the rods.

OPTION 2: CREATE A DECORATED PANEL
MAKE THE SUPPORTS & STEMS

1. Cut the verticals (F) to length.
2. Mark the interior crossbars at 5" and 10".
3. Center the supports over the marks and tack weld in place. Check for square and complete the welds. Grind the welds smooth.
4. Lay out the stems (G) in the desired pattern across the supports.

MAKE THE LEAVES

1. Cut the leaves (H) to desired size using a tin snips or plasma cutter.
2. Place a leaf lengthwise in a bench vise and slightly bend it to form a crease down the center.
3. Arrange the leaves in clusters of three. Weld the leaves together at the base (see photo below, left).
4. Place the leaf clusters on the stem ends and weld in place.
5. Lay out the three panels parallel to each other, hinges touching. Weld each pair of hinges together (see photo below, right).
6. Grind down any rough welds and grind off any spatter. Wipe down with denatured alcohol to remove grinding dust. Finish as desired.

Arrange the leaves in sets of three. Weld the points together, then weld the stem to the leaf sets.

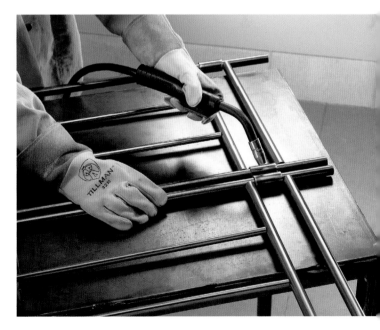

Align the hinge barrels and weld them together.

PART	NAME	DIMENSIONS	QUANTITY
A	Vertical base	⅛ × 1½" flat bar × 30"	1
B	Horizontal bases	⅛ × 1½" flat bar × 10"	2
C	Vertical top	½" hammered bar × 29"	1
D	Horizontal tops	½" hammered bar × 9"	2
E	Rosette	2"	1
F	Scroll set	10 × 10"	1
G	Finials	5¼ × 2⅜"	4

IRON CROSS

Rusted iron cross wall hangings evoke a sense of history. They create a distinctive air of mystery when hung over a fireplace or against exposed brickwork. Design centers and antique stores want hundreds of dollars for something that you can make for much less, so run wild with your imagination on this project and create a cross that personally resonates with you. The more ornate, the better. Search the catalogs from the suppliers listed in the Resource section on page 236 for just the right touches—you'll find dozens of decorative pieces to make this project uniquely yours.

Rather than purchasing a 10- or 20-foot length of hammered or decorative bar—that's a lot of crosses—buy two 39" hammered railing pickets. To get a natural rust finish, thoroughly clean the completed cross with denatured alcohol, wet it down with salt water, and allow it to sit outside for a few months. You'll end up with a naturally beautiful rust.

MATERIALS

- ⅛ × 1½" flat bar (4½')
- ½" hammered bar (4½', Decorative Iron PM-013)
- 2" rosette (1, Decorative Iron LRD-65)
- Scroll set (1, Architectural Iron Designs #55/9)
- Finials (4, Architectural Iron #81/3)

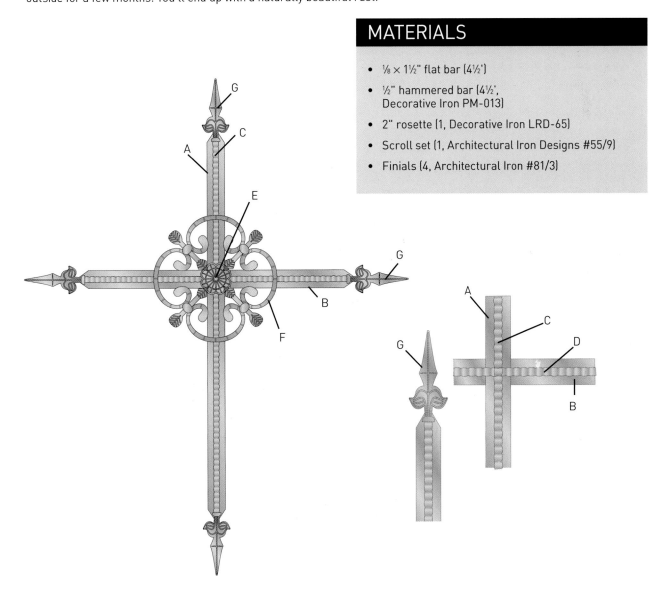

HOW TO BUILD AN IRON CROSS

Before welding, thoroughly clean all parts with denatured alcohol.

MAKE THE BASE

1. Cut the base bars (A and B) to length.
2. Mark the center point of each end of the vertical base and one end of each horizontal base. Mark the bars 1" in from the ends. Draw lines from the 1" mark to the center point to create a point (see photo, below). Cut along the lines.
3. Mark the vertical base 11" from one end. Center the two horizontal bases at this mark to form the cross. Check for square, and weld in place. Grind the welds smooth.

MAKE THE TOP

1. Cut the vertical and horizontal tops (C and D) to length.
2. Center the top vertical bar on the base. The points should extend ½" to ¾" beyond the bar ends. Weld the top bar to the base with two to three welds along each side (see photo, opposite top).
3. Center the top horizontals on the base. They should touch the top vertical bar at the center and end about an inch before the points. Weld in place.

ADD THE DECORATIVE ELEMENTS

1. Center the rosette (E) in the middle of the cross and weld in place.
2. Center the scroll set (F) over the rosette. Bend the arrow tips upward until the scroll set makes contact with all four arms. Weld in place.
3. Center the finials (G) on the bar ends and weld in place.

FINISH THE CROSS

1. Drill ³⁄₁₆" mounting holes through the vertical base on either side of the top bar and about 1" from the top and bottom.
2. Wire brush or sandblast the cross. Finish as desired.
3. Because of the weight of the cross, make sure it is mounted on a wall stud. Hang the cross with 1½" screws.

Mark the midpoint of both vertical base ends and one end of each horizontal base. Mark 1" from the ends. Connect the marks at a diagonal to create a point.

Center the vertical top on the vertical base and weld in place with two to three welds per side.

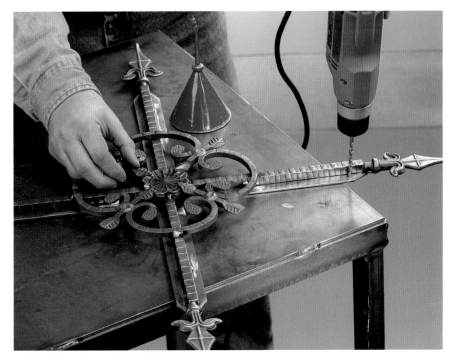

Drill mounting holes through the vertical base. Use machine oil and a slow drill speed to preserve the drill bit.

NESTING TABLES

Do you like to entertain but find that you're constantly searching for extra tables as the guests arrive? Nesting tables are the perfect way to always be ready to entertain on the fly. Best of all, they can be tucked away at the end of the evening. The basic pattern here is for three tabletops. These tables have wooden tops with a ½" overhang. You may adjust the measurements for tile or metal tops. If you want the tables to nest even closer, use ½" angle iron with a 16-gauge or ⅛" metal top. And don't forget to dress these tables up! Consider using hammered bar or adding decorative elements to the legs to allow them to show off a bit at parties.

MATERIALS

- ½ × ½" square bar (28')
- 1 × 1 × ⅛" angle iron (17')
- ½" handipanel or veneered plywood (2' × 4' sheet)
- ¾" #8 panhead screws (12)

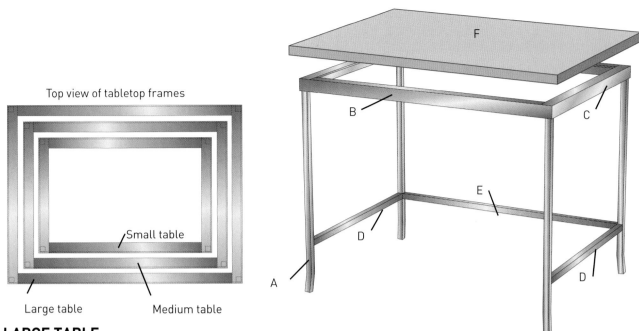

Top view of tabletop frames

Small table

Large table   Medium table

LARGE TABLE

PART	NAME	DIMENSIONS	QUANTITY
A	Legs	½ × ½" square bar × 18"	4
B	Table long sides	1 × 1 × ⅛" angle iron × 23"	2
C	Table short sides	1 × 1 × ⅛" angle iron × 17"	2
D	Crossbars	½ × ½" square bar × 15¾"	2
E	Long crossbar	½ × ½" square bar × 21¾"	1
F	Tabletop	½" plywood 24 × 18"	1

MEDIUM TABLE

PART	NAME	DIMENSIONS	QUANTITY
A	Legs	½ × ½" square bar × 16"	4
B	Table long sides	1 × 1 × ⅛" angle iron × 20"	2
C	Table short sides	1 × 1 × ⅛" angle iron × 14"	2
D	Crossbars	½ × ½" square bar × 12¾"*	2
E	Long crossbar	½ × ½" square bar × 18¾"*	1
F	Tabletop	½" plywood 21 × 15"	1

SMALL TABLE

PART	NAME	DIMENSIONS	QUANTITY
A	Legs	½ × ½" square bar × 14"	4
B	Table long sides	1 × 1 × ⅛" angle iron × 17"	2
C	Table short sides	1 × 1 × ⅛" angle iron × 11"	2
D	Crossbars	½ × ½" square bar × 9¾"	2
E	Long crossbar	½ × ½" square bar × 15¾"	1
F	Tabletop	½" plywood 18 × 12"	1

* Approximate dimensions, cut to fit.

HOW TO BUILD NESTING TABLES

Using a carpenter's square, lay out the table sides. Tack weld together and check for square before final welding.

For a wood or metal tabletop, weld the legs inside the corner formed by the flanges. Align the curved legs so they point toward the long sides.

Before welding, thoroughly clean all parts with denatured alcohol.

MAKE THE LEGS

1. Cut the legs of the large table (A) to length.
2. Clamp the lower $1\frac{1}{2}$" of a leg into a sturdy bench vise. Slide a 3' length of $\frac{1}{2}$" pipe over the leg. Bend it about 5°.
3. Using the first leg as a template, bend all the remaining legs to match.
4. Repeat steps 1 to 3 with the medium and small tables.

MAKE THE TABLETOP FRAMES

1. Cut the table long sides (B) and table short sides (C) to length. Miter the ends at 45°, or cut notches in the long sides.
2. Lay out one long side and one short side of the large table using a carpenter's square to check for square. Clamp into place and tack weld the corner (see photo, top left).
3. Repeat Step 2 with the other long and short sides.
4. Use a carpenter's square to align the two Ls to form a rectangle.
5. Weld the outside joints together. Grind down the welds.
6. Repeat steps 1 to 5 with the medium and small tables.

ATTACH THE LEGS

If you will be making a wooden or metal tabletop, turn the table frame so the flanges of the angle iron are pointing downward. Weld the legs inside the corner (see photo, bottom left). If you will be making a tile tabletop, turn the frame so the flanges are pointing up to create a cradle for the base and tile. Weld the legs to the bottom of the frame (see photo, opposite top).

1. Place a leg on or in a corner, with the bend facing the long side. Use a carpenter's square to check that it is square from front to back and side to side. Weld in place.
2. Repeat step 1 with the remaining legs.
3. Measure the distance between the legs on the short sides, close to the frame. Cut the crossbars (D) to fit. Weld the crossbars into place at 4" from the bottom of the legs.
4. Measure the distance between the legs on a long side, close to the frame. Cut the back bar (E) to fit. Weld into place, even with the crossbars.
5. Repeat steps 1 to 4 with the medium and small tables.
6. Grind down all the welds. Sand, wire brush, or sandblast the tables. Finish as desired. (If using a metal top, install it before finishing.)

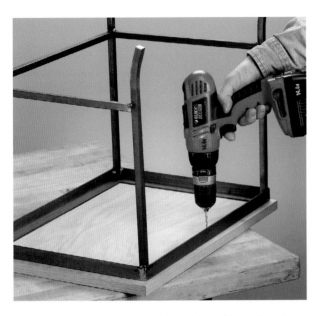

After drilling holes in each of the table's sides, align the wood top and drill pilot holes for the mounting screws.

For a tiled tabletop, weld the legs to the flat underside so that the flanges form a cradle for the tile and backing material.

OPTION 1: CREATE A WOOD TOP

The wood tabletops have a ½" overhang.

1. Cut the wood tabletop (F) to size.
2. If using a handipanel, sand the edges smooth, slightly rounding the top edge and corners, or use a ½" roundover bit and rout the edges.
3. If the top is veneered plywood, apply edge banding with an iron. Trim the banding to fit and lightly sand the edges and corners.
4. Apply paint, stain, or finish.
5. Turn the table upside down and drill a ³/₁₆" hole in each side. Drill a corresponding ⅛" pilot hole in the underside of the tabletop (see photo, top left). Attach the top to the table with ¾" panhead screws.
6. Repeat steps 1 to 5 with the medium and small tables.

OPTION 2: CREATE A TILE TOP

1. Measure the inside dimension of the tabletop. Cut ¾" exterior-grade plywood to fit.
2. Mark the plywood piece with your desired tile layout.
3. Apply a small amount of silicone adhesive to the lip of the table. Place the plywood onto the table.
4. Apply a layer of adhesive to the back of each tile using a notched trowel. Place the tiles on the plywood according to your pattern. After placing the tiles, use a carpet-covered 2 × 4 and a rubber mallet to set the tiles.
5. After the adhesive sets, apply grout to the tile spaces according to manufacturer's instructions. Do not apply grout to the gap between the metal frame and the tiles. Use colored silicone sealant between the metal and the tiles.
6. Repeat steps 1 to 5 with the medium and small tables.

OPTION 3: CREATE A METAL TOP

You can make a metal top of matching steel or create a contrasting look with expanded sheet metal, patterned stainless steel, or aluminum. A glass panel covering patterned metal creates a smooth surface while allowing the pattern to shine through.

1. Cut the metal to fit the dimensions of the table.
2. Tack weld the top to the table from the underside, or weld around the entire perimeter. Grind the contact points to create a smooth transition.
3. Repeat steps 1 and 2 with the medium and small tables.

SWING-AWAY COAT HOOKS

These good-looking, easy-to-make coat hooks are considerably sturdier than those you might purchase at discount stores. The pivoting design lets them fold flat against the wall when not in use, so you can install them wherever you want.

Customize the hooks with a special shaped back plate or use cutout letters to personalize hooks for a child's room. Make the hooks shorter, if desired, or use more than three hooks. Mount the hooks on a stud to prevent the weight of coats from pulling them out of the wall.

MATERIALS

- 16-gauge ½" round tube (9")
- ¼" round rod (10")
- ³⁄₁₆" round rod (6½')
- ⅛ × 2" flat bar (12")
- 1" O.D. ball ¼" 20 threaded (2, Decorative Iron #C50035)

PART	NAME	DIMENSIONS	QUANTITY
A	Hooks	³⁄₁₆" round rod × 9"*	3
B	Crossbars	³⁄₁₆" round rod × 7"*	3
C	Scrolls	³⁄₁₆" round rod × 9"	3
D	Barrels	16-gauge ½" round tube × 3"	3
E	Pin	¼" round rod × 10"	1
F	Balls	1" round balls	2
G	Back plate	⅛ × 2" flat bar × 12"	1

*Approximate dimensions. Cut to fit.

HOW TO BUILD SWING-AWAY COAT HOOKS

Before welding, thoroughly clean all parts with denatured alcohol.

MAKE THE SCROLLS

1. Cut the scroll blanks (C) to length. Round the rod ends with a grinder.
2. Clamp one end of the scroll blank to a $\frac{1}{2}$" pipe. Bend the scroll around the pipe a three-quarter turn.
3. Clamp the other end of the scroll to a 1" pipe. Bend the scroll around the pipe a three-quarter turn in the opposite direction.
4. Repeat with the other two scrolls, matching the bends in the first scroll.

MAKE THE HOOKS

1. Cut the hooks and crossbars (A and B) to length. Round one hook end with a grinder.
2. Place 1" of the hook into a bench vise and bend to about 110°. Repeat with the other two hooks.
3. Align a crossbar at 90° to the hook and weld in place (see photo, bottom right).
4. Trim the hook and crossbar ends to match, if necessary.
5. Place a scroll between the legs of a hook. The hook legs should be no more than $2\frac{3}{4}$" apart.
6. Arrange the other hooks and scrolls to match the first. Weld together.

MAKE THE HINGE

1. Cut the barrels (D) to length. Draw a guideline down the center of each barrel. Make a mark at $\frac{1}{2}$" down from the top of each barrel along the guideline.
2. Align the hook with the barrel so the upper leg of the hook is at the intersection of the guideline and the $\frac{1}{2}$" mark.
3. Weld the hook legs to the barrel.
4. Repeat steps 2 and 3 with the other two hooks.
5. Cut the pin (E) to length. Grind down $\frac{1}{2}$" at each end of the pin to fit the holes in the balls. Slide the barrels over the pin. Place the balls (F) over the pin at each end.

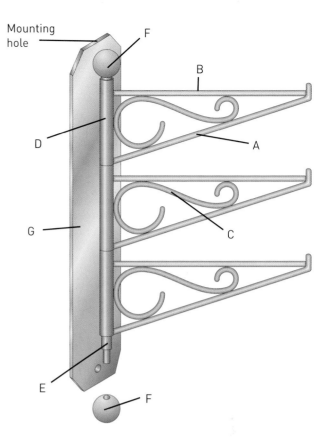

ASSEMBLE THE COAT HOOKS

1. Cut the back plate (G) to size. Round the corners with a grinder.
2. Drill $\frac{5}{32}$" mounting holes centered 1" from each end.
3. Measure the distance between the centers of the balls on the hook assembly. Drill $\frac{1}{4}$" holes this distance apart on the centerline of the back plate.
4. Place the hook assembly on the back so the balls rest in the holes. From the back side, weld the balls to the back plate.
5. Sand, wire brush, or sandblast the coat hooks. Finish as desired.

Align the crossbar with the hook and weld the ends together. A magnetic clamp helps hold the small parts in place.

NIGHTSTAND

Sleek and sophisticated, this table may be designed like a wood table, but its unmatched style comes from metal. Many people find metal easier to work with than wood, and metal can offer the same great looks without all that sawdust.

When working with thinner materials, make sure to turn your welder to the appropriate setting to prevent burn through and distortion.

MATERIALS

- 16-gauge 1 × 1" square tube (14½')
- 16-gauge sheet metal (3 × 5½' sheet)
- 16-gauge angle iron (3')
- ⅛ × ½" flat bar (14')
- 22-gauge sheet metal (2 × 1' sheet)
- Drawer handle

PART	NAME	DIMENSIONS	QUANTITY
A	Legs	16-gauge 1 × 1" square tube × 22"	4
B	Table frames	16-gauge 1 × 1" square tube × 20"	4
C	Shelf	16-gauge sheet metal × 14 × 20"*	1
D	Drawer box	22-gauge sheet metal × 18 × 6"	1
E	Drawer glides	16-gauge ½ × ½" angle iron × 18"	2
F	Drawer face	16-gauge sheet metal × 6 × 12*	1
G	Sides	16-gauge sheet metal × 18 × 6"	2
H	Back	16-gauge sheet metal × 12 × 6"	1
I	Bars	⅛ × ½" flat bar × 11"*	13
J	Tabletop	16-gauge sheet metal × 20 × 20"*	1
K	Drawer handle		1

*Approximate dimensions, cut to fit.

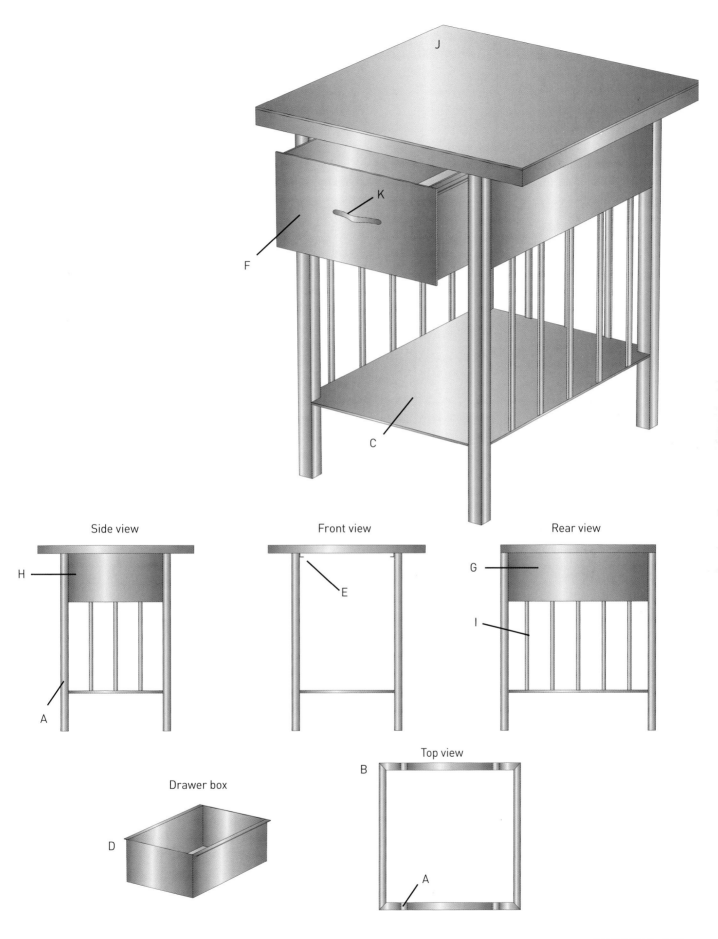

Side view

H

A

Front view

E

Rear view

G

I

Drawer box

D

Top view

B

A

Clamp a leg in place with a piece of angle iron and two C-clamps. Check for square and weld the leg to the table frame.

HOW TO BUILD A NIGHTSTAND

Before welding, thoroughly clean all parts with denatured alcohol.

MAKE THE TABLE FRAME

1. Cut the legs (A) and table frame (B) to size. Miter the frame ends at 45°.
2. On a flat work surface, position two sides of the frame in an L shape. Use a carpenter's square to check for square and then clamp in place. Tack weld the corner. Repeat with the other two sides.
3. Arrange the two Ls to make a square. Clamp in place. Measure across the diagonals to check for square. If the diagonal measurements are not equal, the assembly is not square. Adjust and recheck. Tack weld the corners.
4. After rechecking for square, finish welding all the joints.
5. Grind down the welds to be flush with the surrounding metal.
6. Mark points 3" in from the ends on one frame side. Mark points 3" in from the ends on the opposite frame side.
7. Align a leg with the mark, and tack weld into place. Use clamps and an angle iron to securely hold the leg (see photo, left). Repeat with the other three legs.
8. Make sure the legs are perpendicular to the frame and weld in place.

MAKE THE SHELF

1. Before cutting the shelf (C) to size, measure between the outside faces of the front and side legs to make sure this dimension is correct. Cut the shelf to size.
2. Make 1 × 1" cutouts in each corner of the shelf using a plasma cutter, oxyacetylene torch, or bandsaw. Note that the square tube has slightly rounded edges, so round the cutouts to match.
3. Mark each leg at 5" from the bottom.
4. Align the shelf with the marks, and tack weld in place. Make sure the shelf is level before finish welding.

MAKE THE DRAWER

1. Cut the drawer (D) to size.
2. Use a hand seamer to bend the drawer sides at ⅜" to 90°.
3. Bend the sides and ends of the drawer to 90° to make a box (see photo, opposite top).
4. Weld the seams.

FINISH THE DRAWER

1. Measure the distance between the two front legs. Subtract ⅛" and cut the drawer face (F) to size.
2. With the drawer in the frame, place the drawer face over the drawer front. Align the drawer face with a 1⁄16" gap between the frame and legs. Placing magnetic right-angle clamps in the drawer makes this task easier.
3. Tack weld the drawer face to the drawer from the inside.
4. Remove the drawer and complete the welds.
5. Center your chosen drawer pull on the drawer face and mark the screw locations. Drill the holes for the pull.

ADD THE SIDES & BACK

1. Before cutting the sides (G) and back (H), measure between the side legs and the back legs to determine the exact lengths. Cut the back and sides to size.
2. Place the sides and back, and weld in place.
3. Measure from the top of the shelf to the bottom of the sides and back to determine the correct length for the bars (I). Cut the bars to length.
4. Place three bars evenly spaced (approximately 3" on center) across the back. Make sure the bars are perfectly vertical before welding. Place five bars evenly spaced (approximately 3" on center) across each side (see photo, middle right). Make sure the bars are perfectly vertical before welding.

FINISH THE TABLE

1. Measure the top outside dimension of the table frame. Cut the tabletop (J) to size using these dimensions.
2. Attach the tabletop from the underside using two small welds per side. If you prefer a tabletop that appears one with the frame, weld around the entire perimeter of the tabletop and grind the weld smooth.
3. Wire brush, sand, or sandblast the table and drawer. Finish as desired.
4. Attach the drawer pull.

MAKE THE DRAWER GLIDES

1. Cut the glides (E) to size.
2. Mark ¼" down from the top inside of each leg. Align the glides with the lines and tack weld in place (see photo, bottom right).
3. Slide the drawer into the frame and check for fit. Adjust if necessary, and complete the welds.

Cut the drawer to size. Clamp the drawer to a sharp-edged surface using C clamps and a piece of angle iron. Bend the sides, back, and front of the drawer to 90°.

Weld the decorative bars to the table sides and shelf every 3". Weld from the inside.

Weld the drawer glides to the legs.

FIREPLACE CANDELABRUM

Warm up your hearth, without all the chopping and stacking of wood, with this wavy candelabrum. The simple waves imitate tongues of flames, while the large candleholders support illuminating pillar candles. Romantic, simple, and great for warm summer evenings when a hot fire might be too much.

MATERIALS

- ⅛" sheet metal (13 × 19")
- 4½" bobeches (3, Architectural Iron Designs #79/12)

PART	NAME	DIMENSIONS	QUANTITY
A	Waves	⅛" sheet metal × 4 × 19"	5
B	Bobeches	4½"	3

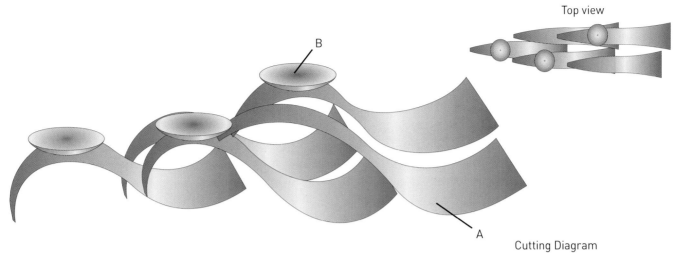

Top view

Cutting Diagram

HOW TO BUILD A FIREPLACE CANDELABRUM

Before welding, thoroughly clean all parts with denatured alcohol.

CUT THE WAVES

1. Mark the sheet metal at 4", 4½", 8½", and 9" along one 13" side, and 2", 2½", 6½", 7", 11", and 11½" along the opposite 13" side.
2. Use a straightedge and a permanent marker or soapstone to connect the marks (see Cutting Diagram).
3. Cut the waves (A) to size using an oxyacetylene torch or plasma cutter.
4. Grind the cut edges smooth, if necessary, and sand or wire brush the metal.

BEND THE WAVES

1. Clamp the narrow end of a wave to a 4" bending form. Bend the wave around just slightly past halfway (see photo).
2. Clamp the wide end of the wave to a 4" bending form. Wrap the wave around the tube a half turn in the opposite direction from the first wave.
3. Repeat steps 1 and 2 with the other four waves.

ARRANGE & FINISH

1. Arrange the waves as desired. Make sure the arrangement fits in the fireplace.
2. Weld the waves together at the contact points.
3. Place the bobeches on the waves, making sure they are level. Drill a ¼" hole in the center of each bobeche.
4. Weld the bobeches to the waves by welding through the hole. If you choose to use brass bobeches, braze or braze weld the bobeches to the waves.
5. Finish the candelabrum as desired.

Clamp the narrow wave end to a 4" bending form and bend it slightly past halfway.

MAGAZINE RACK

Form follows function with this nifty M design. It's a magazine rack that's a snap to put together and perfect for keeping magazines organized and off the floor. You can use virtually any combination of sheet metal and tubing to make this rack. In this project, 16-gauge sheet metal and ¾ × 1½" rectangular tubing is used. You could also try it with round tubing, hammered tubing, pierced sheet metal, or expanded sheet metal. Make it to fit your favorite magazine, and it will be truly customized.

Note: This is a sharp-edged project and not recommended for households with small children.

MATERIALS

- 16-gauge sheet metal (20 × 12" sheet)
- 16-gauge ¾ × 1½" rectangular tube (40")

PART	NAME	DIMENSIONS	QUANTITY
A	Sides	16-gauge sheet metal 10 × 12"	2
B	Legs	16-gauge ¾ × 1½" rectangular tube × 10"	4

HOW TO BUILD A
MAGAZINE RACK

Before welding, thoroughly clean all parts with denatured alcohol.

ASSEMBLE THE SIDES

1. Cut the sides (A) to size.
2. Remove slag and grind down any cutting imperfections.
3. Set up the sides in an upsidedown V. The V angle should be between 50° and 60°. Clamp the sides in place and tack weld the sides together (see photo).
4. Weld the sides together with four 1" welds.

CUT THE LEGS

1. Cut the legs (B) to length.
2. To get the exact angle measurement for the angle at the top of the legs, trace the outside of the V on a piece of paper. Align a carpenter's square along the tops of the V, with one side of the V contacting the corner of the square. Mark the line for the leg (as though you are making an M). Repeat for the other side.
3. Flip the carpenter's square over and align it with one of the newly drawn lines. Align a leg in the corner of the square. Using a straightedge, mark the leg where it crosses the V. Repeat on the same side. Mark the remaining two legs using the other side of the pattern.
4. Cut the leg angles.

ASSEMBLE THE RACK

1. Place the V upside down. Place a leg about ½" in from the front or back end of the V. Use a try square to check that the leg is perpendicular, and tack weld in place.
2. Repeat step 1 with the other three legs.
3. Turn the rack upright to check for leg alignment. Make adjustments, if necessary, and complete the welds.
4. Grind down welds and spatter. Wire brush the rack and wipe down with denatured alcohol before finishing.

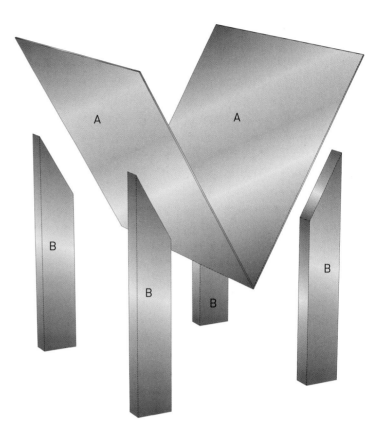

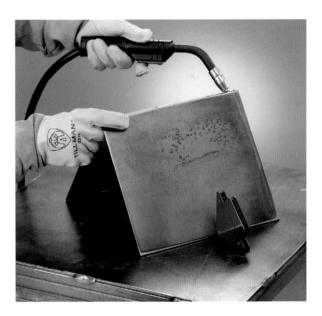

Form a V with rack sides and clamp with magnetic clamps. Weld the seam.

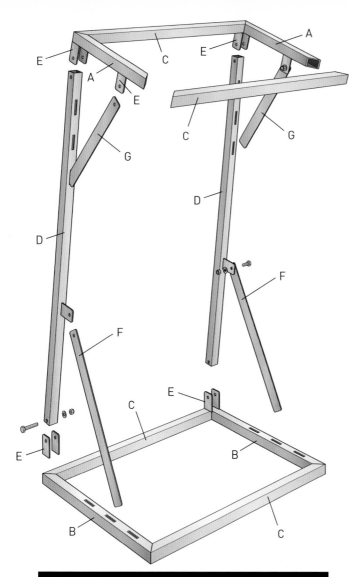

FOLDING LAPTOP / MEAL TABLE

After a long day at work or a fun evening in the shop creating, settle in with a custom-made laptop/meal table. Work or play, this table will allow you to work in comfort in a cozy chair or enjoy that Sunday afternoon snack during the big game. Lightweight and adjustable, it's made just for you!

MATERIALS

- 1 × 1" 11-gauge steel square tubing (16')
- ³⁄₁₆ × 1¼" HRS flat bar (78")
- ¼ × 2" bolts (4)
- ¼" flat washers (16)
- ¼" locknuts (8)

PART	DIMENSIONS	STOCK	QUANTITY
A	12"	1 × 1" 11-gauge steel square tubing	2
B	18"	1 × 1" 11-gauge steel square tubing	2
C	24"	1 × 1" 11-gauge steel square tubing	4
D	36"	1 × 1" 11-gauge steel square tubing	2
E	2"	³⁄₁₆ × 1¼" HRS flat bar	10
F	16"	³⁄₁₆ × 1¼" HRS flat bar	2
G	12"	³⁄₁₆ × 1¼" HRS flat bar	2

HOW TO BUILD A FOLDING LAPTOP/ MEAL TABLE

BUILD THE BASE FRAME

Cut two pieces of part B, and two pieces of part C, and then bevel-cut each end of all pieces to 45 degrees.

1. Lay out the pieces to form an 18 × 24" rectangle, and check to make sure the frame is square. Tack in place, and then measure again to ensure the assembly is square (measure diagonals: if the measurements are equal, the frame is square).
2. If necessary, adjust the frame for squareness, and then weld all around, in four places.

CREATE THE WORK PLATE FORM

1. Cut the two pieces of part A and two pieces of part C, and bevel-cut each end to 45°.
2. Lay out the pieces to form a 12 × 24" rectangle, and check to make sure the frame is square. Tack in place, and then measure again to ensure the assembly is square (measure diagonals: if the measurements are equal, the frame is square).
3. If necessary, adjust the frame for squareness and then weld all around, in four places.

CREATE THE VERTICAL LEG

1. Cut 10 pieces of part E, and two pieces of part D. On one end of part E, round the corners and drill a ⁵/₁₆" hole ½" in from the end and centered.
2. On one end of each tube, (D), measure and mark a point 12" from the end. Center and weld one E tab to this location on each vertical leg.
3. Align and tack remaining eight tabs (E) in place (see photo, bottom right).
4. On one end of F and G, drill a ⁵/₁₆" hole ½" in from the ends, and centered. Round off the corners.

ASSEMBLE THE TABLE

1. Along the 18" leg of the base frame, mark a center line along the tubing, and then measure, space, and mark three ¼ × 1½" slots, 2" apart.
2. Repeat the above steps for the vertical legs.
3. Attach the vertical legs to the base frame and work platform with ¼" bolts. Attach the 12" adjustment arms to the tabs welded to the work platform. Attach the 16" adjustment arms to the tabs welded to the vertical legs.

Weld flat bar support tabs in place.

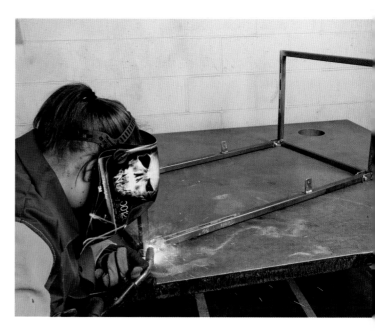

Tack weld the part E tabs in place.

ANGLED METAL SHELVES

This shelf design is very versatile. The pattern can be altered infinitely—change the width, height, depth, and number of shelves to create a unit to fit in any space and display any item. Dress it up or down by using different materials. Consider using shelves with smooth tubing, a metallic finish, and glass (as we have done here), or use hammered tube, an antique finish, and weathered wood shelves for a rustic look.

MATERIALS

- ¾ × 1½" rectangular tube (24½')
- ⅛ × ¾" flat bar (2½')
- Tempered glass or wood (6')

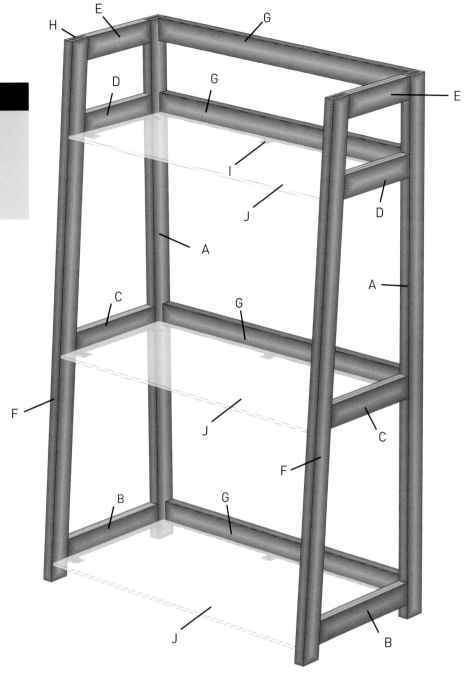

PART	NAME	DIMENSIONS	QUANTITY
A	Back legs	16-gauge ¾ × 1½" rectangular tube × 36"	2
B	Bottom shelf sides	16-gauge ¾ × 1½" rectangular tube × 9"	2
C	Middle shelf sides	16-gauge ¾ × 1½" rectangular tube × 7¾"	2
D	Top shelf sides	16-gauge ¾ × 1½" rectangular tube × 6½"	2
E	Top sides	16-gauge ¾ × 1½" rectangular tube × 6"	2
F	Front legs	16-gauge ¾ × 1½" rectangular tube × 37½"	2
G	Backs	16-gauge ¾ × 1½" rectangular tube × 22"	4
H	End caps	⅛ × ¾" flat bar × 1½"*	8
I	Shelf supports	⅛ × ¾" flat bar × ¾"	21
J	Shelves	Tempered glass*	3

* Cut to fit.

HOW TO BUILD ANGLED METAL SHELVES

Before welding, thoroughly clean all parts with denatured alcohol.

MAKE THE SIDES

1. Cut the back legs (A) to length.
2. Cut the shelf and top sides (B, C, D, E) to length with one end mitered at 83°.
3. Mark the back legs at 1", 7", and 15".
4. Align the flat end of the bottom, middle, and top shelf side pieces with the marks. Align the top side piece with the top of the back leg. Before welding, hold a straightedge against the angled side to check for alignment. Adjust if necessary. Tack weld in place. Repeat with the other back leg and sides.

5. Align a front leg (F) with the side and back leg assembly. Using a carpenter's square held against the back leg, mark the top and bottom angled cuts for the front leg (see photo, below). Repeat with the other front leg.
6. Cut the first front leg to length along the marks.
7. Tack weld the front leg to the side. Repeat with the second front leg.

To get the exact angle for cutting the front leg, align the tube with the side pieces and use a carpenter's square to mark a line even with the top of the top side.

After assembling the sides, weld the backs in place. Check for square both vertically and horizontally.

Weld the shelf supports to the bottom of the sides and backs.

ASSEMBLE THE SHELVES

1. Cut the backs (G) to size.
2. Stand a side unit on its back and line up a back piece with the top of the unit. Use a carpenter's square to check that the back is square to the side. Tack weld in place (see photo above, left).
3. Align the second side unit with the back piece and tack weld in place.
4. Repeat with the other three back pieces, aligning them with the top, middle, and bottom side pieces. Check for square and complete all welds.
5. Cut the end caps (H) to fit the tops and bottoms of the front and back legs. (The front pieces will need to be longer due to the angle.) Weld the caps to the leg ends. Grind down all the welds.

INSTALL SHELF SUPPORTS & FINISH

1. Cut the shelf supports (I) to length.
2. Round the end of each with a bench grinder.
3. Weld two supports to the underside of each top, middle, and bottom shelf side piece. Weld three supports to each back piece (see photo above, right). Weld two supports 1" in from the sides, and center the third on each back piece.
4. Sand, sandblast, or wire brush the unit. Finish as desired.
5. Measure the dimensions for each shelf and order tempered glass to fit, or cut wooden shelves.

WALL-MOUNTED SHELF

This wall-mounted shelf is perfect for displaying your decorative jars that hold cooking oils or spices. A single 24" square tile used for ceilings is cut to size at 24 × 6", which gives the appearance of four separate 6 × 6" tiles. A variety of tin tile sources are available on the Internet: see Resources on page 236 for addresses. You may also find ceiling tiles at salvage stores and antique stores—and they already have the antique character. The tiles don't need to be welding-quality steel; installing them with two-part epoxy is sufficient.

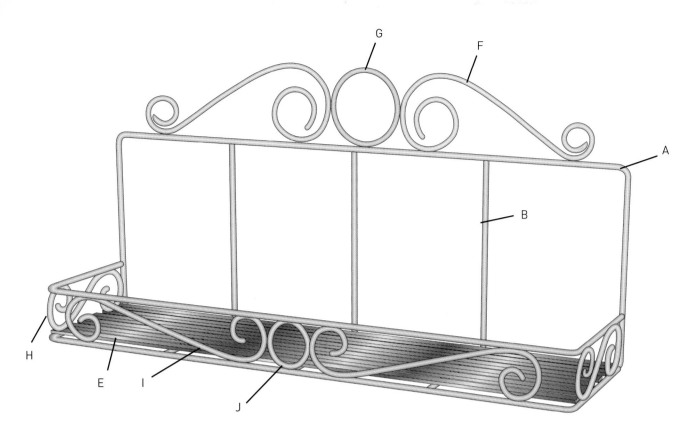

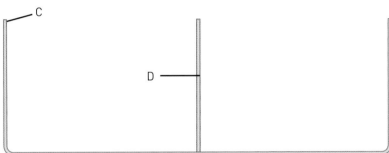

Bottom shelf

MATERIALS

- ¼" round rod (18½')
- ⅛ × ½" flat bar (12')
- 24" metal ceiling tile (1, M-BOSS #0613)
- ³⁄₁₆" round rod (5')

PART	NAME	DIMENSIONS	QUANTITY
A	Frame	¼" round rod × 64"	1
B	Tile supports	¼" round rod × 8"	3
C	Shelf frames	¼" round rod × 36"	2
D	Shelf supports	¼" round rod × 6"	1
E	Shelf cross pieces	⅛ × ½" flat bar × 24"	6
F	Top scrolls	¼" round rod × 15"	2
G	Top circle	¼" round rod × 12"	1
H	Side shelf scrolls	³⁄₁₆" round rod × 9"	2
I	Front shelf scroll	³⁄₁₆" round rod × 15"	2
J	Shelf circle	³⁄₁₆" round rod × 10"	1
K	Tin tile	24 × 6"	1

HOW TO BUILD A WALL-MOUNTED SHELF

Before welding, thoroughly clean all parts with denatured alcohol.

CREATE THE FRAME

1. Cut the frame (A) to length.
2. Mark the frame at 12", 19¾", 43½", and 52¼". Create a rectangle by making 90° bends at the marks. Weld the ends together.
3. Cut the tile supports (B) to length.
4. Weld the tile supports into the frame at 6" intervals (see photo, below).

CREATE THE SHELF

1. Cut the shelf frames (C) to length.
2. Mark each of the shelf frames at 6" in from each end. Make 90° bends at the marks to create the shelf frames.

3. Weld one shelf frame to the base of the frame at a 90° angle.
4. Cut the shelf supports (D) and cross pieces (E) to length.
5. Weld the first shelf support to the installed shelf frame at the midpoint. Weld the remaining two supports centered between an outside edge of the shelf frame and the center shelf support.
6. Space the shelf cross pieces at ½" intervals across the supports. Weld in place.
7. Weld the second shelf frame to the frame 2" above the shelf at a 90° angle (see photo, opposite left).

Weld the tile supports to the frame at 6" intervals.

Weld the top shelf frame to the frame 2" above the shelf.

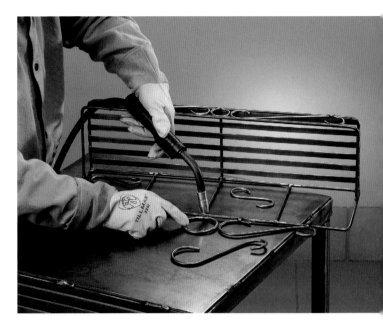

Weld the decorative scrolls and circles to the frames.

MAKE THE SCROLLS

1. Cut the top (F), side (H), and front (I) scrolls to length.
2. Clamp the top scroll rod to a 2" pipe. Bend the rod one full turn around the pipe. Clamp the other end around a 1" pipe and bend the rod one full turn around the pipe in the opposite direction from the first bend. Repeat with the second top scroll, making sure the scrolls match.
3. Clamp the end of a front shelf scroll to a 1" pipe. Bend the rod one full turn around the pipe. Clamp the other end to the pipe and bend the rod one full turn around the pipe in the opposite direction. Repeat with the second front scroll.
4. Clamp the end of a side scroll to a 1" pipe. Bend the rod one turn around the pipe. Clamp the other end to a 1" pipe and bend the rod one turn around the pipe in the opposite direction. Repeat with the second side scroll.
5. Bend the top circle (G) around a 2" pipe, overlapping the ends. Cut the overlap off and weld the circle ends together.

6. Bend the shelf circle (J) around a 1" pipe, overlapping the ends. Cut off the overlap and weld the circle ends together.
7. Place the shelf unit on its back and align the top scrolls and circle along the top. Weld at the contact points (see photo above, right).
8. Align the front scrolls and circle in between the shelf and the shelf support. Weld at the contact points. Repeat with each of the side scrolls.

FINISH THE SHELF

1. Cut the tile down to a 6 × 24" strip, or 6" squares, depending on the tile and pattern you have chosen.
2. Use two-part epoxy to glue the tiles to the frame.
3. Sand, wire brush, or sandblast the shelf.
4. Finish as desired.

ÉTAGÈRE

Étagère is a French term for a metal structure used to display items in the house or plants in the garden. Étagères can be shockingly expensive, but this project is not only affordable, it is much sturdier than many available in stores. The corner étagère that follows (on page 162) is designed to match the style of this project.

This decorated shelf may be used inside or outside the home. Paint it or allow it to gently rust. The decorative elements are not structural, so you may change them to suit your desired level of fuss. The flat expanded sheet metal was chosen because it repeats the diamond pattern, but anything from metal to wood to tile can be used as shelving material. Check out the decorative iron suppliers listed in the Resources section on page 236 for dozens of friezes and metal stampings that may be used instead of the diagonals and scrolls.

MATERIALS

- 16-gauge 1 × 1" square tube (23')
- 16-gauge ½" flat expanded sheet metal (2 × 3')
- ¼" round rod (31')
- ⅛ × ½" flat bar (5')
- 1½" balls (4, Triple S Steel #SF116F4)
- 1 × 1" × ⅛" angle iron (6')

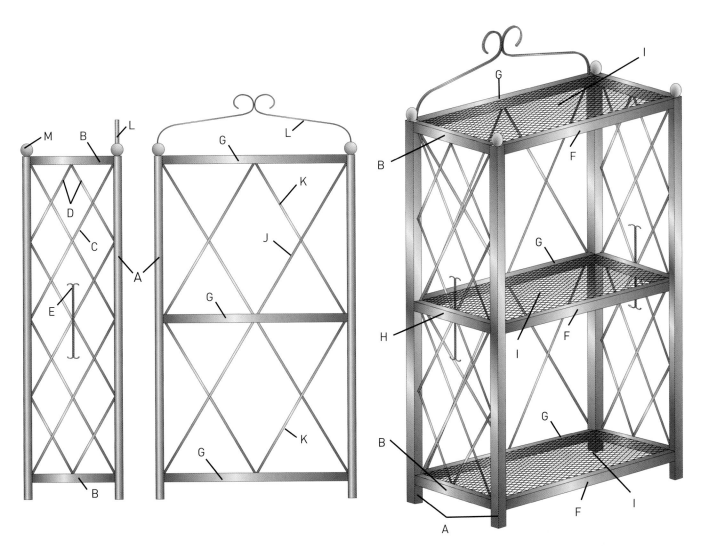

PART	NAME	DIMENSIONS	QUANTITY
A	Legs	16-gauge 1 × 1" square tube × 40"	4
B	Crossbars	16-gauge 1 × 1" square tube × 10"	4
C	Long diagonals	¼" round rod × 21"*	12
D	Short diagonals	¼" round rod × 10"*	8
E	Side decorations	⅛ × ½" flat bar × 5¾"	8
F	Front supports	1 × 1" × square tube × 22"	3
G	Back supports	1 × 1 × ⅛" angle iron × 22"	3
H	Shelf supports	⅛ × ½" flat bar × 9"	2
I	Shelves	Expanded sheet metal 23 × 11"	3
J	Back long diagonals	¼" round rod × 43"	2
K	Back short diagonals	¼" round rod × 21"	4
L	Scrolls	⅛ × ½" flat bar × 24"	2
M	Balls	1½" diameter	4

*Approximate dimension, cut to fit.

HOW TO BUILD AN ÉTAGÈRE

Before welding, thoroughly clean all parts with denatured alcohol.

MAKE THE SIDES

1. Cut the legs (A) and crossbars (B) to length.
2. Mark each leg 2" from one end. This is the bottom of the leg.
3. On a flat work surface, lay out two legs and two crossbars. Align the bottom of one crossbar with the marks. Align the top of the other crossbar with the top of the legs. Clamp in place.
4. Check all corners for square. Tack weld all corners. Recheck for square by measuring across the diagonals. If the two measurements are equal, the unit is square. If the measurements are not equal, make adjustments until they are. Complete the welds.
5. Repeat steps 3 and 4 to make the second side panel.

CUT THE DECORATIVE SIDE DIAGONALS

1. Find and mark the midpoint of the top and bottom crossbars.
2. Find and mark the midpoint of the front and back legs between the crossbars.
3. Mark the points halfway between the midpoints and the crossbars. These are the quarterpoints.
4. Determine the lengths for the long diagonals (C) by measuring from a corner to the opposite midpoint and from an upper quarterpoint to a lower, opposite quarterpoint (see photo, below). These measurements should be equal if the panel is square. Cut the long diagonals slightly longer than this measurement.
5. Determine the lengths for the short diagonals (D) by measuring from a quarterpoint to the adjoining crossbar midpoint. Cut the short diagonals slightly longer than this measurement.
6. Hold the diagonals in place against the side panels and mark the necessary cutting angle. Cut at the marks.

Measure between a corner and the opposite midpoint to find the length for the long diagonals.

Weld the first layer of diagonals in place.

ASSEMBLE THE DECORATIVE SIDE DIAGONALS

1. With the side panel on a flat surface, lay the three long and two short diagonals that slant down from upper left to lower right in place. Weld to the frame (see photo, above).
2. Lay the three long and two short diagonals that slant from upper right to lower left in place on top of the other set of diagonals. Weld in place.
3. Cut the side decorations (E) to length. Round one end of each to a smooth semicircle using a bench grinder.
4. Clamp the rounded end to a ½" pipe and bend the bar a half turn around the tube. Repeat with all the pieces.
5. Place two pieces back to back and weld together. Repeat with the other pieces.
6. Grind the flat ends of the decoration to a V.
7. Place one welded piece vertically at the middle of the diagonal pattern and weld in place. Place the second welded piece upside down at the middle and weld in place (see photo, page 160, top left).
8. Cut and assemble the diagonals for the second side panel.

ASSEMBLE THE ÉTAGÈRE

1. Cut the front supports (F) and back supports (G) to length.
2. The decorations on the side panels should be flush on one side and recessed on the other. The flush side is the outside. Place the two sides parallel to each other lying on their backs. Align the front supports with the top and bottom crossbars and centered over the midpoint. Check for square and tack weld in place (see photo, page 160, top right).
3. Orient the top back support so the flat faces are to the top and front and aligned with the top crossbar. Tack weld in place.
4. Orient the midpoint and bottom back supports with the flat faces to the top and back. Align with the front of the back leg. (This will allow the back panel decorations to be recessed).
5. Cut the shelf supports (H) to length.

(CONTINUED)

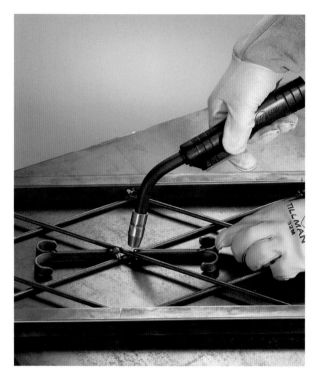

Weld the decorations to the middle × of the diagonals.

Tack weld the front supports in place and check for square.

6. Weld the shelf supports flush with the top of the middle shelf frames at the ends. Bend the top shelf to fit inside the four supports (see photo, top left, on page 161). Weld in place.
7. Cut the shelves (I) to size.
8. Bend the front of the middle shelf to fit behind the front support. The shelf lies on top of the shelf suport and back support. Weld in place.
9. Bend the front and sides of the bottom shelf to fit between the supports. The back lies on top of the back support. Weld in place.

MAKE THE BACK PANEL DECORATIONS
1. Measure the diagonal across the back of the shelves from corner to corner. Cut the back long diagonals (J) to this length.
2. Measure from the midpoint of the top back shelf support to the midpoint of the leg panel. Cut the short diagonals (K) to this length.
3. Place the left to right descending diagonals in place and weld. Place the right to left diagonals and weld.

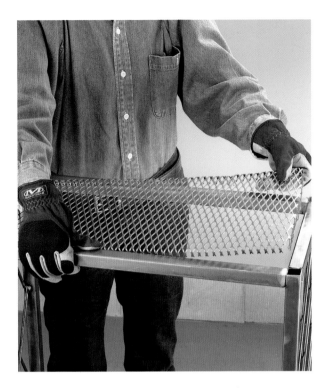

Bend the top shelf to fit inside the sides and supports.

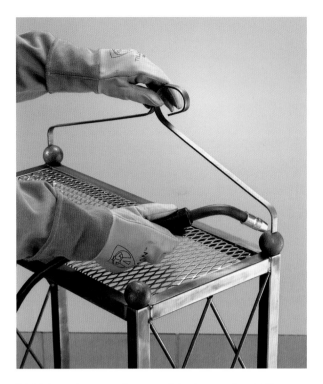

Weld the spheres to the tops of the legs and weld the scroll to the tops of the rear spheres.

MAKE THE DECORATIVE SCROLLS

1. Cut the scrolls (L) to length.
2. Round one end of each scroll to a smooth semicircle using a bench grinder.
3. Clamp the flat end to a 4" pipe, and bend slightly past a one-quarter turn to create a rounded 90° angle. Repeat with the second scroll.
4. Cut a ⅛" notch in a 2" pipe.
5. Insert the rounded end of the scroll through both notches. Bend the scroll three-quarter turn around the pipe. Repeat with the second scroll.

6. Weld the balls (M) into the tops of the legs (see Tip, below).
7. Align the two scrolls so they meet in the middle and are symmetrical. Weld at the contact points.
8. Weld the scroll to the top of the rear balls (see photo above, right).
9. Sand, wire brush, or sandblast the shelf unit. Finish as desired.

TIP

WELDING THICK MATERIAL TO THIN MATERIAL

Welding solid spheres to 16-gauge material won't be successful without an extra step. The spheres will not weld easily because they absorb a great deal of heat before melting. Heating them first with a propane torch assures a good weld.

CORNER ÉTAGÈRE

Why have an empty corner when it can display a beautiful shelf? Designed to match the étagère on page 156, this shelf unit is perfect for displaying plants or other decorative curios. You don't need to hide it in a corner—it can stand on its own in your backyard, or you can build two to create a semi-circle shelf to stand out along a wall.

The arcing shelf fronts look fancy but are actually fairly simple to execute. Face the front with a decorative frieze or metal stampings for a different look. The diagonals do not provide structural support, so you may personalize the sides in any way you choose.

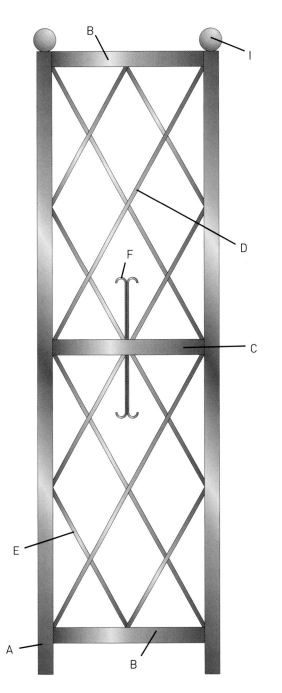

MATERIALS

- 16-gauge 1 × 1" square tube (10')
- ¼" round rod (15½')
- ⅛ × 1" flat bar (4½')
- 16-gauge ½" flat expanded sheet metal (2 × 2')
- 1½" balls (3, Triple S Steel #SF116F4)
- 1 × 1 × ⅛" angle iron (20")

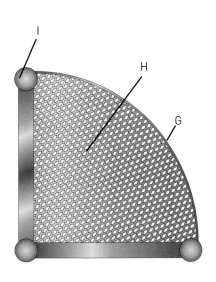

PART	NAME	DIMENSIONS	QUANTITY
A	Legs	16-gauge 1 × 1" square tube × 40"	3
B	Crossbars	16-gauge 1 × 1" angle iron × 10"	4
C	Middle crossbars	1 × 1 × ⅛" angle iron × 10"	2
D	Long diagonals	¼" round rod × 21"*	12
E	Short diagonals	¼" round rod × 10"*	8
F	Decorations	½ × ⅛" flat bar × 5¾"	8
G	Shelf fronts	⅛ × 1" flat bar × 18"	3
H	Shelves	16-gauge expanded sheet metal*	3
I	Balls	1½" dia.	3

* Cut to fit.

HOW TO BUILD A CORNER ÉTAGÈRE

Before welding, thoroughly clean all parts with denatured alcohol.

MAKE THE FRAME

1. Cut the legs (A) and crossbars (B) to length.
2. Measure 2" up from the ends of the legs and make a mark.
3. Align one crossbar in between and flush with the tops of two legs. Check for square and tack weld in place. Align the bottom of a second crossbar with the end marks, check for square, and tack weld in place.
4. Measure across the diagonals of this unit to check for square. The measurements should be equal; if not, adjust until they are. Weld in place.
5. Repeat steps 3 to 4, attaching the third leg and the crosspieces to the leg unit at right angles to form a triangular frame (see photo below, left).
6. Cut the middle crossbars (C) to length, mitering one end at 45°. Center the crossbars over the midpoints and even with the insides of the legs. Weld in place.

CUT THE DECORATIVE DIAGONALS

1. Find and mark the midpoint of the top and bottom crossbars.
2. Find and mark the midpoint of the front and back legs between the crossbars.
3. Mark the points halfway between the midpoints and the crossbars. These are the quarterpoints.
4. Determine the lengths for the long diagonals (D) by measuring from a corner to the opposite midpoint and from an upper quarterpoint to a lower, opposite quarterpoint. Cut the long diagonals to length. Determine the lengths for the short diagonals (E) by measuring from a quarterpoint to the adjoining crossbar midpoint. Cut the short diagonals to length. Place the diagonals and mark the angled cut necessary for an exact fit (see photo below, right).

Attach the third leg and crosspieces to create a 90° angle.

After cutting the diagonals to approximate length, mark the angles where the diagonal crosses the framework.

ASSEMBLE THE DECORATIVE DIAGONALS

1. With the side panel on a flat surface, lay the three long and two short diagonals that slant down from upper left to lower right in place. Weld to the frame.
2. Lay the three long and two short diagonals that slant from upper right to lower left in place on top of the other set of diagonals. Weld in place.
3. Cut the side decorations (F) to length. Round one end of each to a smooth semicircle, using a bench grinder.
4. Clamp the rounded end to a ½" round tube and bend the bar a half turn around the tube. Repeat with all the pieces.
5. Place two pieces back to back and weld together. Repeat with the other pieces.
6. Place one welded piece vertically upright at the middle X of the diagonal pattern and weld in place. Place the second welded piece upside down at the middle X and weld in place.
7. Cut and assemble the diagonals for the second side panel.

MAKE THE SHELVES

1. Cut the shelf fronts (G) to length.
2. Bend a shelf front into a 12" radius arc by making small bends every ½". Check that it fits in between the two front legs. Adjust, if necessary. Bend the other two shelf fronts.
3. Weld the shelf fronts to the forward inner corners of the front legs (see photo below, left).
4. Cut the shelves (H) to size. Allow ¾" bending allowance on all sides.
5. Use a pliers and a hand seamer to bend the shelf to fit inside the top and bottom crossbars and shelf fronts. The middle shelf rests on top of the crossbars. Cut notches in the sheet metal, if necessary, to help it fit (see photo below, right).
6. Weld the shelves in place.

FINISH

1. Place balls (I) in the tops of each leg and weld in place (see Tip, page 161).
2. Sand, wire brush, or sandblast the shelf unit. Finish as desired.

Weld the arched fronts to the sides, even with the crossbars.

Use a hand seamer and pliers to bend the expanded sheet metal to fit inside the top and bottom crossbars.

OUTDOOR LIFE

TRUCK RACK

A carrying rack that mounts on the walls of your truck bed is a very handy accessory for serious metalworkers and DIYers with a frequent need to transport building materials. The heavy-duty metal rack shown here is made from 12-gauge steel tubing and can be rescaled to fit onto just about any pickup truck, from compact to full-size. The full-length bedrails distribute the load, protect the truck's bedrail, and provide rack attachment points along the entire length of the truck bed. They also provide a flat base for mounting a toolbox or cover. If your truck has stake holes and you prefer to align the uprights to fit into those spaces, you must alter the rack design a bit—this is a valid design consideration for ease of installation and less drilling into your truck.

Perhaps the most unique feature of this rack is the round spacer plates that fit between the upper rack bars. The circular cutouts are very handy for tying down a load.

To protect the rear window of the cab, consider attaching a panel of expanded metal mesh to span between the front uprights. You can weld the mesh directly to the uprights or build a metal sub-frame to fit between the

uprights and weld the mesh to the subframe. Also consider mesh for over the cab to protect the truck, as we have done.

Even when scaled for a compact pickup, this rack will weigh at least 200 pounds. The final weight, of course, depends on the size of the rack and the gauge of the metal you use. In any event, you will need at least one helper to lift the rack into position. A person for each corner is best. Finally, be sure to finish the metal so that it does not rust.

MATERIALS

- 12-gauge ($\frac{1}{8}$") 2 × 2" steel tubing, approx. [(dim. a + dim. c) × 4] + (dim. b × 4)

- $\frac{1}{8}$ × 3 × 3" angle iron (dim. a. × 2)

- $\frac{1}{4}$ × 4 × 4" flat stock steel (5$\frac{1}{2}$')

- $\frac{1}{4}$ × 2 × 2" flat stock steel

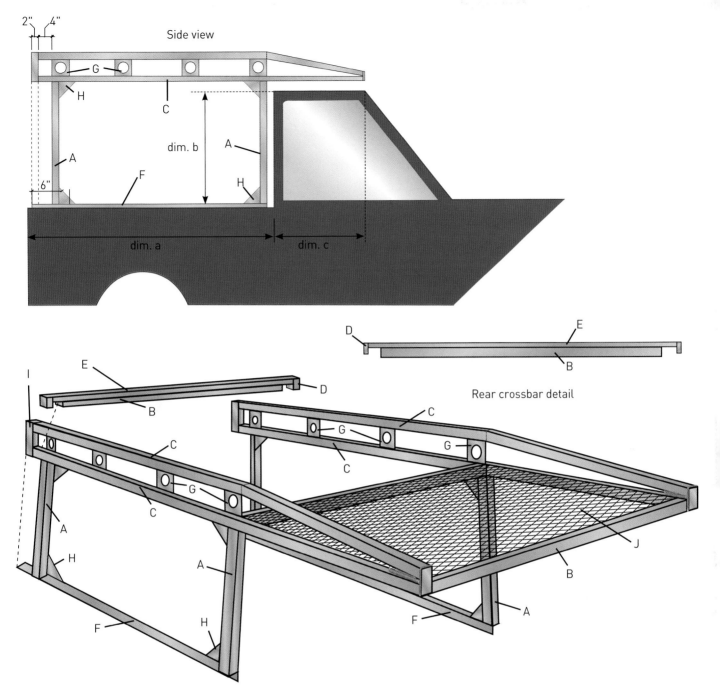

Side view

2" 4"

G

H

C

dim. b

A

A

F

6"

H

dim. a

dim. c

D

E

B

Rear crossbar detail

E

B

D

I

C

G

C

G

A

C

G

C

H

A

A

F

H

J

B

A

F

PART	NAME	DIMENSIONS	QUANTITY
A	Uprights	$\frac{1}{8}$" × 12-gauge 2 × 2" steel tubing × (dim. b + 3")	4
B	Crossbars	$\frac{1}{8}$" × 12-gauge 2 × 2" steel tubing × (dim. d – 4")	3
C	Long top bars	$\frac{1}{8}$" × 12-gauge 2 × 2" steel tubing × (dim. a + dim. c)	4
D	Rear crossbar tabs	$\frac{1}{4}$" × 2" flat stock steel, 2$\frac{1}{2}$"	2
E	Rear crossbar top	$\frac{1}{4}$" × 2" flat stock steel	1
F	Angle iron on bedrails	$\frac{1}{8}$" × 3 × 3" angle iron × (dim. a)	2
G	Spacer plates with holes	$\frac{1}{4}$" × 4" × 4' flat stock steel 4 × 4"	8
H	Gussets	$\frac{1}{4}$" × 4" × 4' flat stock steel 4 × 4" hypotenuse	8
I	End piece	2 × 2 steel tubing × 8"	2

* Caps in sizes 2 × 2" and 2 × 4" needed for all exposed tube ends.

HOW TO BUILD A TRUCK RACK

MEASURE THE TRUCK

1. Measure corner to corner on your truck bed for square (see photo, top, right).
2. Measure and mark your truck bedrails for angle iron (F) placement. The full length of the truck bed is dimension a. This dimension is required to design your rack so that it custom fits your truck.
3. Measure the length of your truck cab roof from front to back (dimension c)—refer to drawing on page 169. Add this measurement to dimension a for the total length of the long top bars (C).
4. Measure from the truck bedrail to the top of the cab. Add 1" for additional clearance above the cab. The height from the truck bedrail to the top of the cab is dimension b. *Note: The uprights on this rack lean toward the center at a 10° angle so that the long top bars (C) are aligned roughly with the inside walls of the truck bed.*

PREPARE THE ANGLE IRONS

1. Mark two $\frac{1}{8} \times 3 \times 3$" angle irons (F) to length (dimension a). Use a speed square to ensure a straight cutting line across the parts.
2. Cut the two angle irons at marks using a cutoff saw. *Option: If you choose not to bolt the angle iron to the truck and instead use clamping all along the bedrail angle iron or some other fastening technique, skip to Cut & Fasten Uprights to Angle Irons.*
3. Align the angle irons on the truck bedrails. Align the solid edge butted up flush against the inside edge of the truck bed so it caps over the bedrails, and clamp in place.
4. Make a mark 6" in from the tailgate edge of the angle irons. The 2 × 2 uprights are aligned 6" in from the tailgate edge. The cab-side upright sits flush to the ends of the angle irons (F) on the cab-side edge, so make a mark 2" in from the cab-side edge of the angle iron as well (see photo, bottom, right).
5. Measure in from the marks made in step 4 for four to five evenly spaced bolts along the angle iron (F). It is standard to use four bolts to fasten a truck rack to a smaller truck; larger trucks will require more bolts. Check with your truck manufacturer and the fastener manufacturer's recommendations.
6. Remove the angle irons and bring them back to your shop.
7. Slowly drill $\frac{5}{16}$" holes through the angle iron (F) for grade 8 $\frac{3}{4}$" hex bolts at your bolt marks using a drill press or handheld drill. Use cutting oil, and drill slowly. Repeat for the other angle iron (F).

Measure from corner to corner of the truck bed for square.

Measure in 6" from the tailgate ends of the angle iron, and make a mark for the tailgate uprights. Make a mark 2" in from the cab-side ends of the angle irons for cab-side uprights.

CUT & FASTEN UPRIGHTS TO ANGLE IRONS

1. On a flat surface, measure and mark 12-gauge ($\frac{1}{8}$") × 2 × 2" steel tubing lengths at dimension b for the four uprights (A). This will leave a 4" clearance between the rack and your truck cab roof.
2. Mark for a 10° angle cut on each upright end (see photos of assembled rack on pages 172 and 173). This angle cut ensures the rack fits roughly parallel with the cab roof.

SAFETY

NOTICE: For safe use of this rack project only use approved, safety-rated products to attach the rack securely to your vehicle. These may include clamp-based, bolt-based, or stake-pocket systems. Inspect clamps and fasteners regularly to make sure they are tightened properly.

Tack the cab-end uprights (A) onto the angle iron bedrails (F). Then tack the tailgate-side uprights onto the angle iron bedrails 6" in from the edge. Tack only along two sides, because you may need to break your tack welds to realign parts several times.

3. Cut the uprights to length using a cutoff saw. Refine the 10° angle with an angle grinder.
4. Align the cab-side uprights flush with the angle iron (F) cab-side ends. Tack along two sides so that you can easily break the tacks and realign, if necessary. *Note: The location of the cab-side uprights determines where you need to bend the long top bar. The edges of A and G and the bend on C should all be aligned.* Align the tailgate-end uprights 6" in from the angle iron ends (see the illustration on page 169). Tack along two sides so that you can easily break the tacks and realign, if necessary (see photo, above).

BUILD THE BASIC FRAME

1. Measure and mark the two lower long bars (C) to the length of the truck bed (minus 2") plus the length of the cab (dim. a – 2") + dim. c. *Note: When planning out the four long top bar lengths, remember to account for end caps. There is a 2 × 2" tubing cap at the tailgate end and ¼" flat stock steel cap at the front end. Some racks hang slightly over the cab front while others (such as our rack) sit behind the cab front. The length of your rack depends on your truck size and how much length you need for typical loads. Consult with your truck manufacturer on recommended clearances.*
2. Measure and mark the two top long bars (C) for length (dim. a + dim. c). *Note: The bend itself creates a variable in the final length of the top bars. When the long top bar bends to meet the lower long bar, the two parts do not meet flush at the ends. The square end is created with a ¼" flat stock steel end cap(s); also, the small triangular space in between the two parts is filled with slag.*
3. Cut all four long top bars to length using a cutoff saw.

4. Align the two lower long bars flush on top of the uprights. The long bars will extend beyond the tailgate upright by 4". *Note: The top and lower long bars will be joined with a 2 × 2" steel tubing (8" long) cap at the tailgate end. To have the top bars align with the tailgate end of your truck, the top bars must be aligned 2" in from the tailgate end.*
5. Tack the lower long bars to the uprights. Make minimal tacks so that you can break the welds and realign parts, if necessary. *Note: Consider adding a temporary brace made of scrap steel tubing that spans from the tailgate ends of the top bars to help maintain the proper angle. We realigned the top frame later and then introduced the brace, but it may save time to add the brace now to avoid making time-consuming breaks and re-tacking later.*
6. Measure between the front two edges of the long top bars. Also measure between the long top bars at the cab-side (in between two uprights). Use these measurements to mark the lengths of the two permanent crossbars (B).
7. Cut the two crossbars to length with a cutoff saw.
8. Align the crossbars in between the two lower long bars and then tack them to the long bars. Use four 1" welds evenly spaced 1 to 2"; do not weld all the way down the joint.
9. Measure between the two long bars at the tailgate side of the truck for the removable crossbar. Measure, mark, and then cut this crossbar to length using a cutoff saw.
10. Measure and mark the two rear crossbar tabs (D) to 2 × 2½". Also measure and mark the rear crossbar top (E) to length (length of removable crossbar + 2"). Cut the tabs and top to size.
11. Center the crossbar top (E) over the removable crossbar (B) and clamp together with a corner clamp. Tack weld the two parts together.
12. Align the crossbar tabs (D) flush with the top edges of the crossbar top (E) ends, and then weld them together (see Rear crossbar detail on page 169).
13. Measure and mark eight 4 × 4" square spacer plates (G) from ¼ × 4 × 4" flat stock steel.
14. Cut the spacer plates using a portable band saw.
15. Mark the center circle on each spacer plate using a piece of 2½" (o.d.) pipe as a guide.
16. Cut the center circles out of the spacer plates using a plasma cutter (see photo, page 172, top left).
17. Center the spacer closest to the tailgate 5" in from the long bar end and make a mark (C). This leaves a 4" space in between the end piece (I) and spacer for the rear removable crossbar.
18. Align a spacer with its cab-side edge flush with the cab-side edge of the upright. Mark this placement on the long bar (C).

Cut the circle in the center of spacer plates using a plasma cutter. Shown here: clamped 4 × 4" flat stock (with circle cutout premarked) and ground secured to table.

19. Evenly space the other two spacers along the long bar. Mark the placements onto the lower long bars (C).
20. Repeat the alignment of the spacer plates (G) along the other long bar (C).
21. Align each spacer plate on the lower long bars (C) in alignment with your marks.
22. Hold the spacer plates in place with pliers and tack weld each spacer to the long bars.

ADD LONG TOP BARS

1. Align the two uppermost long top bars (C) directly above the lower long top bars. Center them over the spacer plates.
2. Tack weld the uppermost long top bars to the spacer plates every several inches.
3. Mark the bend location flush with the edge of the cab-side spacer plate and upright.
4. Check the entire rack for proper alignment. Break tack welds and realign parts as necessary. Make adjustment marks and re-tack parts together in proper alignment. *Note: We had to break the welds at the lower top bar and uprights to adjust the uprights to proper alignment with 10° angles at each upright.*
5. When the long bars are removed from the uprights (this is very likely going to be necessary; see Note above), take advantage of this situation to make the long top bar bends on a sturdy worktable.
6. Score along the bend mark using an angle grinder or reciprocating saw. Score only as deep as necessary to make the bend.
7. Bend the top long bars to meet the bottom long bars.
8. Mark the cutting line at the front (cab-side) of the uppermost long top bars for the final length. Cut the excess length off the top bars with a portable bandsaw.

Bend the notched top long bar to reach the lower bar. The front finished triangular ends fill in with slag. The ends are capped with 2× flat stock steel. Inset: The top long bar was capped with a 2 × 2" flat stock steel to bring ends flush; then a 2 × 4" flat stock steel piece was used for the final end cap that spans from top to bottom of the two long bars.

Re-attach the top frame to the uprights, welding at the inside corner of each crosspiece and long bar.

9. Tack weld the top long top bar to the lower long top bar at the front only. *Note: Leave the bend open (not welded) for now.*

RE-FASTEN TOP & BOTTOM FRAMES TOGETHER

1. With the long bars tacked, realign the top/lower long top bars (top frame) atop the uprights and check for proper alignment. *Note: We also tacked on a temporary brace made of scrap steel tubing over the tailgate ends of the lower long bars (C) to help maintain alignment (see photo, bottom, right).*

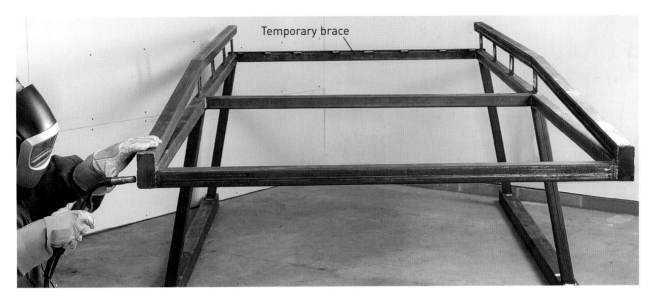

Temporary brace

Weld the final 2 × 4" caps onto the ends of the long top bars for flush front edges.

2. Re-tack the long top bars in place on the uprights. Check for proper alignment and make final adjustments, as necessary.
3. Finish the welds at the bend (previously tacked to allow for adjustments), allowing slag to fill the void.
4. Finish the welds at the fronts of the long bars, allowing slag to fill in between the triangular corners.

ADD END CAPS & GUSSETS

1. Cap the exposed ends of long bar tubing for a flush front edge. *Note: We had to use a 2 × 2" end cap on the long top bar and then another 2 × 4" end cap to cover both the top and lower long top bars.*
2. Cap all remaining exposed tube ends.
3. Cut 12-gauge (⅛") 2 × 2" steel tubing to 8" (I). Align each piece against the lower and uppermost long top bars at the tailgate end of the rack. Weld in place.
4. Measure and mark the eight triangular gussets (H) to fit in four inside corners of the rack (see illustration on page 169).
5. Cut the gussets with a band saw.
6. Align the gusset plates in each inside corner of the rack, and then weld them in place.

FINISH

1. Fully weld all final joints. Make at least 2" welds at the inside corner of each crosspiece and long bar. Do not weld any braces used. And do not weld the rear removable crossbar.
2. Grind down all of the welds (see photo, bottom right) and sand all surfaces in preparation for primer.
3. Break the tacks on any temporary braces and remove them. Grind down those final edges.

4. Prime the entire truck rack to protect it from the elements. Allow the rack to dry.
5. Paint or powder-coat the entire truck rack. Use exterior-rated enamel spray paint. Powder coating is best, but it is also the most expensive and requires special equipment.

INSTALL THE RACK

1. To install the rack onto your truck, you'll need a helper or two. Lift the rack up onto the truck and align the angle irons onto the truck bedrail.
2. Fasten the rack angle irons to the truck with clamps, bolts, and/or your chosen fastening method.

Grind all welds and caps smooth with an angle grinder.

TRAILER FRAME

Most trailers have a very specific purpose. It could be to transport lawn equipment or snow removal machines for a small business, it might be to haul all-terrain vehicles or dirt bikes, or perhaps to transport building materials for DIY projects or cart away trash to the dump on the weekend. Each purpose carries special requirements that are reflected in the trailer construction. The welding project seen here is a highly adaptable trailer frame that can be finished and accessorized in just about any way you can imagine to serve your own needs.

A trailer is basically a metal frame with axles and wheels that you attach with a hitch to your vehicle. When choosing an axle kit, consider weight capacity, type of suspension (leaf spring, torsion bar, etc.), width of frame, and height of the hitch. Axle kits typically include the axle, brackets to fasten the axle to the trailer frame, suspension (ours included leaf springs), fastening system to attach the leaf springs to the axle (ours included U-bolts, plates, bolts, and nuts), and a hub set. Our kit also included the trailer fenders, tires, and wheels—some kits include necessary parts for brakes and lights as well. The axle kit shown in this project was purchased from Northern Tool Company. The cost for a similar axle kit ranges from $170 to $300.

You'll also need to choose and purchase a floor material. For vehicle transport—such as snowmobiles or ATVs—

expanded metal lath is an excellent choice. If you'll be hauling construction debris or yard waste, exterior-rated plywood is a good choice; hardwood boards (such as white oak) are a good option for high durability. Walls can also be formed with similar materials.

Further customization of your trailer can include tie-downs and anchors, removable or hinged ramps, and custom tongue or coupler hardware.

MATERIALS

- 14-gauge 2 × 2" steel tubing
- $^{3}/_{16}$ × 2" HRS flat bar (7')
- $^{1}/_{4}$ × 1¾" HRS flat bar (1')
- Axle kit (incl. leaf springs, U-bolts, U-brackets, axle)
- Hubs and wheels
- Side stakes (2 × 4 treated lumber or metal tubes)
- Trailer sides, front/back, and bed (treated plywood or expanded metal)
- Trailer coupler

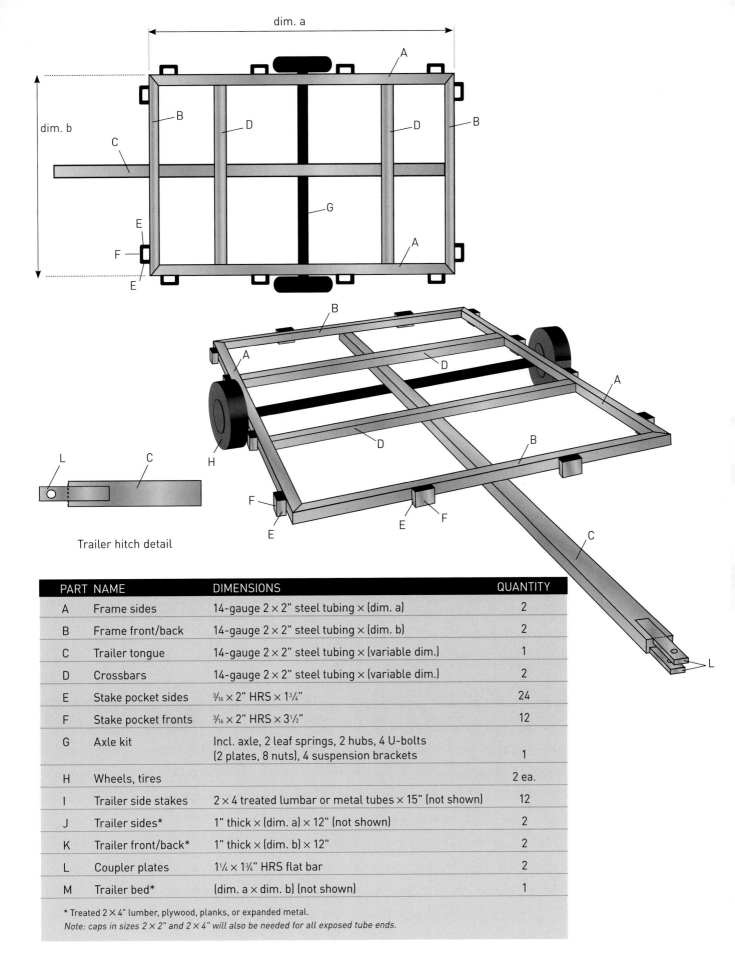

Trailer hitch detail

PART	NAME	DIMENSIONS	QUANTITY
A	Frame sides	14-gauge 2 × 2" steel tubing × (dim. a)	2
B	Frame front/back	14-gauge 2 × 2" steel tubing × (dim. b)	2
C	Trailer tongue	14-gauge 2 × 2" steel tubing × (variable dim.)	1
D	Crossbars	14-gauge 2 × 2" steel tubing × (variable dim.)	2
E	Stake pocket sides	³/₁₆ × 2" HRS × 1³/₄"	24
F	Stake pocket fronts	³/₁₆ × 2" HRS × 3¹/₂"	12
G	Axle kit	Incl. axle, 2 leaf springs, 2 hubs, 4 U-bolts (2 plates, 8 nuts), 4 suspension brackets	1
H	Wheels, tires		2 ea.
I	Trailer side stakes	2 × 4 treated lumbar or metal tubes × 15" (not shown)	12
J	Trailer sides*	1" thick × (dim. a) × 12" (not shown)	2
K	Trailer front/back*	1" thick × (dim. b) × 12"	2
L	Coupler plates	1¹/₄ × 1³/₄" HRS flat bar	2
M	Trailer bed*	(dim. a × dim. b) (not shown)	1

* Treated 2 × 4" lumber, plywood, planks, or expanded metal.
Note: caps in sizes 2 × 2" and 2 × 4" will also be needed for all exposed tube ends.

HOW TO BUILD A TRAILER FRAME

BUILD THE FRAME

1. Measure, mark, and then cut the frame sides (A) and frame front/back (B) parts to length (dimension a and dimension b, respectively). *Note: We cut the ends at a 45° angle for mitered corners.*

2. Square one frame side tube (A) with one frame front tube (B), and tack them together to make an L.

3. Repeat step 2 for the other frame side tube and back tube.

4. Tack the two Ls together and check for square using a combination square or framing square (see photo below, top).

5. Weld the inside corners; then weld the outside corners; lastly, weld along the top and bottom.

6. Cut the two crossbars (D) to length.

7. Align the crossbars in between the frame sides. *Note: The exact location is determined by the axle size and placement. The crossbars add overall support to the frame. Most axle kits come with instructions and placement recommendations based on the size of your trailer and the loads you intend to carry.*

8. Check each crossbar for square, and then tack them onto the frame.

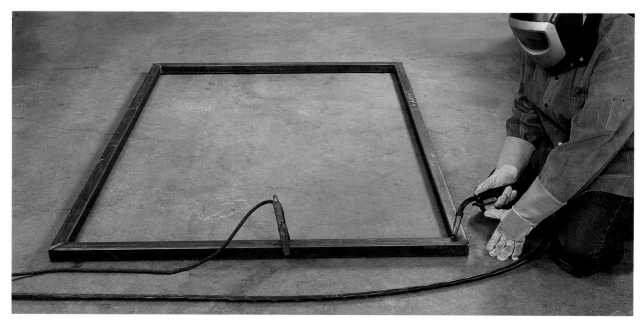

Tack weld together the trailer sides and front/back to make the basic frame.

SAFETY

NOTICE: If you plan on using your trailer on streets or roads it must be licensed and, in most cases, it must have working brake lights, reflectors, and safety chains in case of hitch failure. Some areas may also require that your trailer pass a safety inspection.

Tongue

Weld the trailer tongue to the underside of the frame.

CONNECT THE TRAILER TONGUE TO CROSSBARS

1. Measure, mark, and then cut the trailer tongue (C) to length.
2. Align the trailer tongue down the center of the frame, on top of the crossbars. Clamp the trailer tongue to the crossbars. Measure to ensure the tongue is centered and square. Adjust as necessary. *Note: We extended the trailer tongue to the far back of the trailer frame for optimal strength.*
3. Tack the trailer tongue to the two crossbars, and then weld it onto the frame (see photo, opposite bottom).

INSTALL THE AXLE KIT & WHEELS

1. Align a bracket onto each underside of the trailer frame sides. Follow the axle kit instructions, and make sure the brackets are square and aligned at the exact same location on both sides of the frame.

2. Tack the brackets onto the frame (see photo below, top).
3. Align the leaf spring into the bracket tacked onto the trailer following the manufacturer's instructions. Locate the placement for the other bracket under the leaf spring, and then align it on the trailer frame. Repeat for the other trailer side.
4. Check the brackets on both sides of the trailer for squareness to each other.
5. Set the leaf spring aside, and then weld all brackets in place.
6. Bolt the leaf spring to the brackets following the manufacturer's instructions.
7. Align the axle in place under the leaf spring following the manufacturer's instructions. Fasten the axle to each leaf spring. Our kit came with two heavy-duty U-bolts that wrap around the axle and then bolt onto each leaf spring (see photo below, bottom).
8. Install the hubs and wheels onto the axle following the axle kit instructions. Inflate the tires.

Tack the first axle bracket onto the underside of the trailer frame. Repeat for the other trailer side. The brackets need to be perfectly square to each other. Hold the leaf spring in place over the bracket to align the second bracket on each side. Weld the brackets to the frame, and then bolt the leaf spring to the brackets.

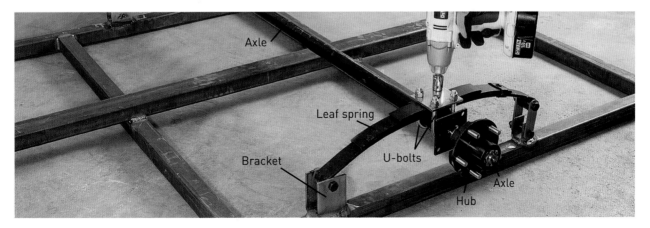

Bolt the axle to the leaf spring with large U-bolts.

MAKE THE TRAILER COUPLER PLATES

1. Cut the coupler plates (L) to size (see photo, below), and then drill holes in the plates. Match the hole size and location to the coupler used (see photo, opposite top). *Note: Our coupler slid over the plates and fastened to the tongue on the top and bottom, so we aligned the plates with holes at top and bottom of the tongue, but some couplers fasten to tongues at the sides and the hole locations will vary. The coupler and hitch type is determined by your vehicle type, final trailer construction, and the weight of your trailer and anticipated loads (see Resources, page 236).*

2. Visually align, clamp, and then tack the coupler plates to the trailer tongue. Extend the coupler plates at least 3" past the trailer tongue end.

3. Mark the angle bend location(s) on the tongue. Determine the tongue angle(s) by measuring from the hitch of the towing vehicle to the trailer tongue (remember to inflate tires). *Note: If the angle is significantly steep, making multiple bends at incremental angles will provide a stronger tongue (see photo, opposite bottom left).*

4. Score the angle mark using a 4½ angle grinder with a fiber cut-off wheel.

5. Bend the tongue to the proper angles.

6. Tack the tongue back together at the scores. Double-check the fit next to the towing vehicle. The trailer should be level (see photo opposite, bottom right).

7. Weld the coupler plates to the tongue.

8. Make all final welds at bends along tongue.

9. Slide the coupler over the coupler plates. Bolt the coupler to the plates, and then weld along all points on which the coupler and tongue connect.

MAKE THE STAKE SIDE POCKETS

1. Measure and cut the two sides for each stake pocket (E) to ³⁄₁₆ × 2" HRS × 1¾".

2. Measure and cut one front for each stake pocket (F) to ³⁄₁₆ × 2" HRS × 3½".

3. Align a pocket side (E) and front (F) onto a 90° magnetic square to form open corner joints (see photo, page 180, top). Tack the pieces together to form an L, and then make the final weld.

4. Use a magnetic square to weld on the second pocket side to form the complete U for all twelve pockets—be sure to use an open corner joint again. Make all final welds.

5. Repeat for all twelve pockets.

6. Align pockets around the trailer so they are evenly spaced along the long frame sides, starting at 2" in from the corners.

7. Tack the pockets to the trailer frame (see photo page 180, bottom). Check for square.

FINISH

1. Roughen the surfaces by sanding.

2. Apply a rusty metal primer for optimal rust protection.

3. Paint or powder coat the trailer with paint rated for exterior applications for extra durability and weatherproofing.

Cut the coupler plates to size using a chop saw.

Drill holes into the coupler plates according to the coupler specifications.

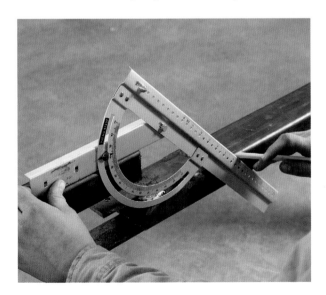

To determine the angle for the trailer tongue, align an angle finder on your truck or ATV hitch and on the trailer tongue. Mark this angle at the appropriate location on the tongue.

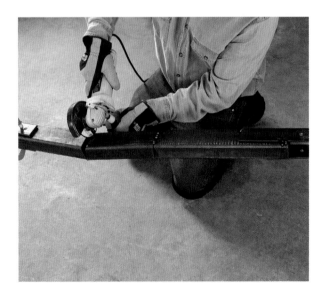

Score notches along the tongue to achieve the desired angle using a handheld grinder. Tack the bends and then hook up the trailer to your towing vehicle—the trailer should be level. If necessary, break tacks and re-bend tongue. Make final welds along each bend.

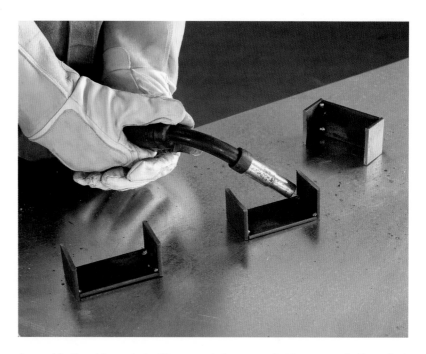

Assemble the side pockets. These pockets are available pre-made if you find this to be too time-consuming. They are relatively inexpensive.

Tack the pockets around the perimeter of the frame.

FLOOR & SIDE OPTIONS

If you plan to use your trailer for yardwork and hauling debris, you'll want to add a solid floor and sides to your trailer. A simple method, which we show here, is to attach plywood siding to 2 × 4 exterior-rated stakes set into the pockets along the trailer frame. The stakes are exterior-treated 2 × 4 lumber and the siding is ¾ × 4 × 8" plywood. The plywood floor is drilled directly to the trailer frame.

MAKE REMOVABLE SIDES & TRAILER BED

1. Measure and cut twelve 2 × 4s (I) into 15" lengths.
2. Measure and cut the plywood bed (M) to dimension a × dimension b.
3. Measure and cut the plywood sides (J) to dimension a × 12".
4. Measure and cut the plywood front/back pieces to dimension b × 12" (h).
5. Insert the twelve 2 × 4 stakes into the trailer pockets.
6. Fasten the plywood sides and front/back to the 2 × 4s (see photo below, top left). If a 2 × 4 does not quite fit, use a chisel and hammer to adjust it.
7. Fasten the trailer bed to the frame (see photo below, top right) with bolts or self-tapping metal screws spaced every 6 to 12".

Fasten ¾"-thick plywood trailer sides to the exterior-rated 2 × 4 stakes.

Drill the exterior-rated plywood sheet for flooring directly to the trailer frame.

Alternative: Weld expanded metal lath all along the top of the frame perimeter (photo left). This is more suitable than wood for hauling ATVs, motorcycles, and snowmobiles. If the primary use of your trailer is to transport vehicles, another feature to consider is a drop-down back (photo right). This requires building out a frame and then welding expanded metal lath onto it—just as you did with the trailer floor—and then using a purchased or custom-made bolt-on hinge system with latches.

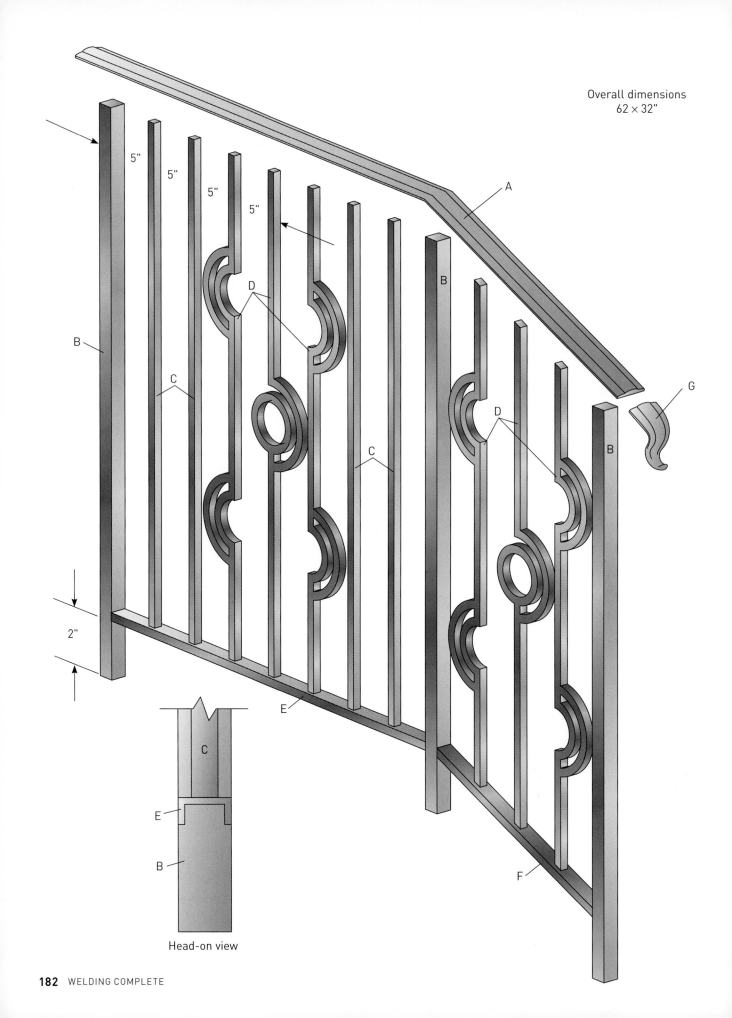

Overall dimensions
62 × 32"

5"

5"

5"

5"

A

B

B

C

D

C

D

B

G

2"

E

C

E

B

F

Head-on view

STAIR RAILING

With the number of companies selling decorative metal pickets and newel posts, it is easy to create a railing distinctly your own. The railing we are making is for a two step concrete stairway.

Some important information about creating railings:

The rail top should not be interrupted by knobs or other decorations—it must allow continuous hand to rail contact and should be between 1½" and 2" in diameter to ensure easy gripping.

Some railings—those mounted against a wall or in an area where the stair surface is not raised above the surrounding surface, such as the stairs in the slope of your lawn—can be a single rail at the appropriate height. A railing that separates the stair, landing, balcony, or deck from a vertical drop has to conform to certain safety standards. Typically, the pickets or balusters must not have any gaps larger than 5½" for child safety. The lower rail should not be more than 2" from the floor surface. If the vertical drop from the floor line is more than 6', the railing must be a minimum of 34" tall. If the vertical drop is less than 6', the railing must be a minimum of 32" in height. Railings must have turn-outs or roundovers at their ends to prevent blunt projections. The safest handrails continue 12" beyond the top and bottom stairs. Railings must be able to withstand a 250-pound force in any direction without giving way. Check your local building codes for specific requirements.

You can purchase rail cap, also called cap rail or handrail, at most steel supply centers. Handrail terminations come in a variety of styles. Those that scroll in an S shape are called "lamb tongues." A "volute" is a spiral termination, and a "lateral" is a flat curl to the side. These may be ordered through specialty railing distributors (see Resources page 236).

MATERIALS

- 1¾" rail cap (as needed)
- 1" square tube (as needed)
- ½" square rod (as needed)
- ½" decorative pickets (as needed)
- ⅛ × ½ × 1" channel (as needed)
- Lamb tongue rail termination
- Mounting hardware

PART	NAME	DIMENSIONS	QUANTITY
A	Handrail	1¾" rail cap × 64"*	1
B	Newel posts	⅛ × 1 × 1" square tube × 32½"	3*
C	Plain pickets	½" square rod × 30"	4*
D	Decorative pickets	½" decorative pickets × 30"	6*
E	Flat bottom rail	⅛ × ½ × 1" channel × 40"*	1
F	Angled bottom rail	⅛ × ½ × 1" channel × 24"*	1
G	Rail termination		1

*Dimensions and quantities must be adjusted to fit the particular stairs.

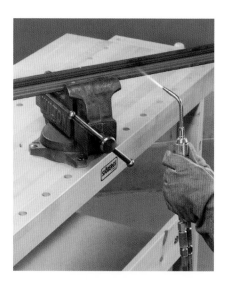

Heat the bending point of the handrail to red hot with an acetylene torch or cutting tip to make the metal easy to bend.

HOW TO BUILD STAIR RAILING

CUT THE HANDRAIL TO LENGTH

1. Measure from the edge of the house to the edge of the landing. Measure from the edge of the landing to where the newel post will be located. Add these two measurements to get the length of the handrail.

2. Cut the handrail (A) to size.

SHAPE THE HANDRAIL

It is a good idea to create an angle guide for bending the handrail by screwing two pieces of wood together at the length and angle to match the stairway.

1. Set the handrail on the landing with one end butting against the house. Mark the handrail at the edge of the landing.

2. Clamp the railing in a bench vise and heat the bending point red hot with an acetylene torch (see photo, left). A cutting torch preheat works best, just make sure you don't hit the oxygen and accidentally cut the metal.

3. When the metal is red hot, bend it to create the angle. It is helpful if you pull on the longer end of the rail to bend so you have more leverage.

4. When finished bending the handrail, place it on the stairs to make sure the bend is correct.

LAY OUT THE NEWEL POSTS & PICKETS

The newel posts will be anchored in the concrete or attached using square footings that will bolt into the concrete, so they have to be far enough from the edge not to destroy the edge of the concrete.

1. Determine the number of plain pickets (C), decorative pickets (D), and newel posts (B) you need.

2. Place the bent handrail on the floor or a large work surface. Place newel posts at each end of the handrail and near the bend.

3. Lay out the pickets in a pleasing pattern, making sure they are no more than 5½" apart (see photo opposite, right).

4. Mark and cut the two bottom rails (E and F) to fit between the newel posts, once you have the layout determined. Place the channel flat side up with the legs down. Mark the picket locations on the railings.

Option: When you look at railings, you will see that some have the pickets and newel posts welded directly to the rail cap, as you see here. Other railings have a piece of channel welded into the underside of the rail cap and the pickets are welded to the flat side of the channel. Punched channel can be purchased with ½ × ½" square holes pierced through it. This is welded under the rail cap, and the pickets are inserted through the punched holes and welded in place. You also can use the punched channel for the bottom rail. Using punched channel means you cannot adjust spacing to account for the unique shapes and sizes of decorative pickets.

CUT THE POSTS & PICKETS TO LENGTH

Purchased decorative pickets range from 36" to 39". When cutting decorative pickets, cut equal amounts from each end unless you wish the pattern to be off center.

1. Determine the height of your handrail and the depth that the newel posts will be footed in the concrete (if they are to be footed; otherwise measure to the top of the concrete).

2. Cut the newel posts and pickets to length. Cut the appropriate angle for the stair pickets.

ASSEMBLE THE RAILING

Concrete can explode when heated, so it is best to do your welds on a sheet of plywood that can be doused with water when you have finished.

1. Tack weld the newel posts to the rail cap. Tack weld the bottom rails to the newel posts.

2. Place the rail assembly on the stairs to make sure the dimensions are correct. If they are not, break or grind off the tack welds and make adjustments.

3. Return the assembly to the work surface, and tack weld the pickets in place, maintaining the proper spacing (see photo below). Use a combination square to check each piece for square before welding.

4. Make the final welds. Weld the lamb's tongue termination to the end of the rail cap.

5. Grind down any rough or unsightly welds. Wire brush or sandblast the rail assembly.

6. Install the railing by cementing the newel posts into the stairs or using bolt-down flange shoes.

7. Prime and paint the railing with a high quality outdoor metal paint.

Lay out the pickets and newel posts in a pleasing arrangement, keeping them no more than 5½" apart, and tack weld the pickets in place maintaining the proper spacing.

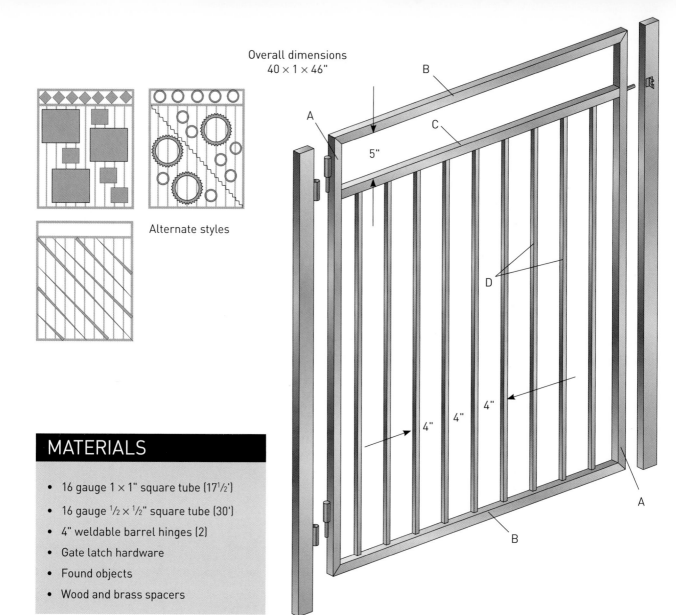

Overall dimensions
40 × 1 × 46"

B

A

C

5"

D

4"

4"

4"

B

A

MATERIALS

- 16 gauge 1 × 1" square tube (17½')
- 16 gauge ½ × ½" square tube (30')
- 4" weldable barrel hinges (2)
- Gate latch hardware
- Found objects
- Wood and brass spacers

Alternate styles

GATE

This gate framework can be used to hold a collection of found metal objects. Use brazing or braze welding to join non-matching metals or thicknesses of metals. Rusty objects will need to be cleaned at the point of contact. If the objects you find are large, you may want to space the uprights at 8", rather than 4". The 4" spacing meets code requirements for gates and fences—this spacing prevents children from getting their heads stuck between uprights.

PART	NAME	DIMENSIONS	QUANTITY
A	Sides	16-gauge 1 × 1" square tube × 46"	2
B	Crosspieces	16-gauge 1 × 1" square tube × 40"	2
C	Interior crosspiece	16-gauge 1 × 1" square tube × 38"	1
D	Vertical inserts	16-gauge ½ × ½" square tube × 40"	9

HOW TO MAKE A GATE

ASSEMBLE THE FRAME

1. Cut the sides and crosspieces (A and B) to size, mitering the ends at 45°.
2. Place the left side piece and bottom crosspiece together at a 90° angle to form one corner of the rectangle. Check the pieces for square, and clamp in place. Tack weld the pieces together.
3. Place the right side piece and top crosspiece together to form another corner of the rectangle. Check for square, clamp in place, and tack weld together.
4. Join the two pieces to form the rectangle. Check the corners for square, and clamp in place. Tack weld the corners together.
5. Measure the diagonals of the rectangle to check for square. If the measurements of both diagonals are equal, the assembly is square. If it is not square, pull or push it into alignment. When aligned, clamp it in place, and finish the corner welds.
6. Turn the assembly over, and complete the welds.

ATTACH THE INTERIOR CROSSPIECE & INSERTS

1. Cut the interior crosspiece (C) and vertical inserts (D) to size.
2. Place the interior crosspiece against the inside edges of the side pieces 5" down from the top crosspiece. Check the pieces for square, and clamp in place. Weld the interior crosspiece to the sides.
3. Place the vertical inserts at 4" intervals between the interior and bottom crosspieces. Make sure the spacing is even—you might need to adjust to slightly less or more than 4" if the miter cuts are slightly off. Keep the outside edge of the inserts flush with the outside edge of the crosspieces.
4. When the spacing is adjusted properly and the inserts are square to the crosspieces, tack weld each insert at both ends.
5. Check the assembly and inserts for square one more time. Turn the assembly over, and weld each upright in place.

ATTACH THE FOUND OBJECTS

Objects other than mild steel will need to be brazed or braze welded. Connect non-metallic objects by wrapping or folding a thin strip of mild steel sheet metal or a short piece of ⅛" steel rod around an edge. Weld the ends of the rod or strap to the framework.

1. Arrange your found objects artfully across the interior space.
2. Carefully clean rust or paint from the areas where the found objects contact the uprights.
3. Weld the found objects in place.

ATTACH THE HINGES

Our gate is made to hang attached to a metal gate post. We chose to use barrel style hinges.

1. Place the gate between the gate posts. Use wood spacers and braces to position the gate between the gate posts, and clamp or brace solidly in place.
2. Line up the hinges on the post and the gate (see photo, below). Use a level to check for plumb. (For the gate to swing properly, the hinges need to be installed perfectly plumb.) Tack weld the hinges to the gate and post.
3. Remove the bracing, and check that the gate swings freely. When it does, complete the hinge welds.
4. Install the gate latch hardware. If the hardware is painted or zinc coated, grind off the coating before welding, or install with screws.

Brace the gate in position between the gate posts. Clamp a wood spacer between the gate post and the gate side. Position the hinge and check for plumb.

OUTDOOR ANGLED FIREPIT WITH FLOOR

Form and functionality will set the mood for your backyard bonfire. Often a firepit can be the centerpiece for an attractive outdoor event area, and this design will allow you to add your own custom touches to an eye-catching design that leaves the normal old, dull round firepit in the scrapyard.

This firepit is a standalone project that will challenge your newfound welding and layout skills. This platform will allow you to tap into your creative side. Add your own decorative patterns to personalize this project. Practice cutting your design on scrap to perfect your plasma-cutting skills.

Work with your local steel provider to shear or plasma cut the 4 × 8-foot sheet into your cut list sizes to allow for safer, easier material handling, especially in a smaller shop. This may add to the cost, but will make for a safer work environment.

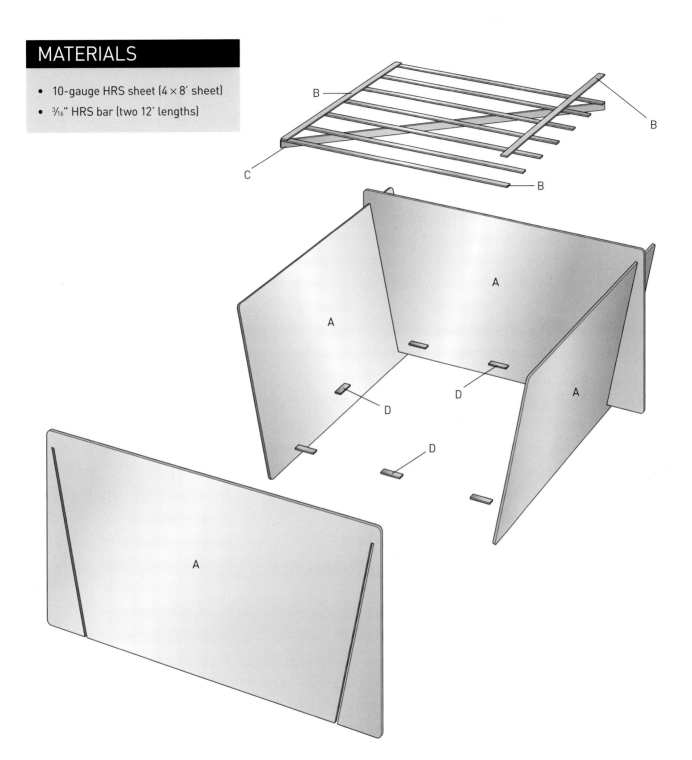

PART	DIMENSIONS	STOCK	QUANTITY
A	24 × 48"	10-gauge HRS sheet	4
B	36"	³⁄₁₆" HRS bar	9
C	51"	³⁄₁₆" HRS bar	1
D	3"	³⁄₁₆" HRS bar	8

HOW TO BUILD YOUR FIREPIT

Before welding, clean all parts to remove mill scale and any oil residue. Wipe clean with denatured alcohol and then grind any remaining mill scale ½" wide through any areas to be welded.

LAY OUT & WELD THE BOX

Place two pieces of part A on the layout table. On one 4' length, measure 6" in from both ends and place a mark. From this point (A), scribe a line at 75° up toward the top outside corner (see photo below, left).

1. From the top of this diagonal line, measure down 2" and in 1", and make a mark—point B. Now, cut along the line from point B to point A, creating a cut equal to the thickness of the 10-gauge sheet, approximately ⅛".

Clean both sides, remove all slag or dross, and set aside the part (see photo below, right).

2. With the second piece of part A, repeat step 1. Using a flap wheel sanding disc, preferably 80-grit, remove all mill scale along this line to a width of 1".

3. Slide part A over the uncut part A; repeat on the opposite side. Flip the assembly over so the bottom side is now face up. Slide parts A along parts B to form a 3 × 3' opening. Measure the diagonals in two directions across the opening to assure it is square (see photo, opposite top left).

4. Tack all four corners, and then measure and stitch-weld the four seams (see photo, opposite top right).

From point A, scribe a line at 75° up toward the top outside corner.

Make a cut along the line from point B to point A.

Measure diagonally to ensure you have a square.

Tack all four corners, and then measure and stitch-weld the four seams.

FABRICATING & INSTALLING
THE FIRE GRATE

1. Lay out four pieces of part B to form a box. Measure diagonally in two directions to make sure the frame is square (see photo below, left). Tack two places at each corner.

2. Position five pieces of part B at 6" on-center intervals across the frame. Measure to ensure both sides are evenly spaced, and then tack the pieces in place. Weld the assembly (see photo below, right).

3. Place and weld the stiffener (C) on its edge along a line diagonally across the underside of the fire grate (see photo, page 192, top).

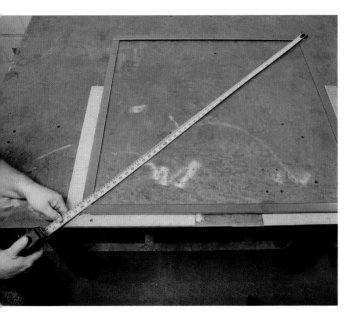

Measure diagonally in two directions to make sure the frame is square.

Position five pieces of part B at 6" on-center intervals across the frame. Measure to ensure both sides are evenly spaced, and then tack the pieces in place. Weld the assembly.

PLACE & WELD THE FIRE GRATE SUPPORT TABS

1. Place one piece of part B in the center along the bottom edge of part A. Weld in place (see photo, right).

2. Place the fire grate in place and mark a line along the underside of the grate on the part B (both sides). Remove grate, and place and tack three pieces of part D evenly spaced along the underside of the scribed line. Repeat on the other side. Weld all pieces in place.

3. Draw your scroll work or decorative design on the sides of the firepit, and then cut them with a plasma cutter or oxyacetylene torch.

Let your creative side be highlighted by the fire on a well-deserved break at night admiring your work!

Place and weld the stiffener on its edge.

Place one piece of part B in the center along the bottom edge of part A. Weld in place.

Remove grate, and place and tack three pieces of part D evenly spaced along the underside of the scribed line. Repeat on the other side. Weld all pieces in place.

Draw your scroll work or decorative design on the sides of the firepit, and then cut them with a plasma cutter or oxyacetylene torch.

VINEYARD TRELLIS

This arbor is a beautiful stand-alone piece, but it is also a brilliant addition to the Arch & Gate (page 198). Either way it is an impressive addition to your yard. Mount it against a wall or fence, or use it as a panel within a fence. Better yet, make an entire fence of trellises! To do so, screw mount the panels to 4 × 4 wood posts, or weld the panels to 3 × 3 steel posts.

You may want to cap off your trellis with a finial or decorate the panels with permanent trailing vines for the off season. See Resources on page 236 for finial, frieze, and stamped metal sources. Make sure to order parts that can be welded. Cast iron cannot be easily welded to the steel trellis parts.

MATERIALS

- 16-gauge 1 × 1" square tube (23')
- 16-gauge $\frac{1}{2}$ × $\frac{1}{2}$" square tube (18')
- $\frac{3}{4}$" round rod (6')
- $\frac{1}{8}$ × $\frac{1}{2}$ × $\frac{1}{2}$" angle iron (4")
- $\frac{1}{4}$" round rod (18')

HOW TO BUILD A VINEYARD TRELLIS

Before welding, clean all parts thoroughly with denatured alcohol.

MAKE THE PANEL

1. Cut the legs (A) and crossbars (B) to length.
2. Lay out the crossbars between the legs. Align the top crossbar with the top of the legs. Place the lower crossbar 4" from the bottom of the legs.
3. Check for square, clamp in place, and tack weld together.
4. Measure across the diagonals to check for square. If the diagonals are not equal, adjust until they are. Complete the welds.
5. Cut the verticals (C) to length.
6. Mark the crossbars at 5", 10", 15", and 20". Center the verticals at the marks, check for square, and weld in place.

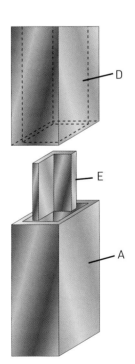

Sleeve detail

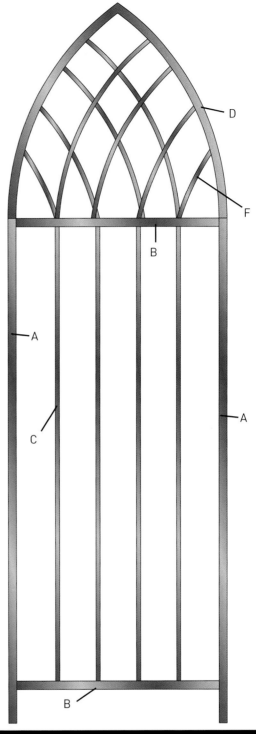

PART	NAME	DIMENSIONS	QUANTITY
A	Legs	16-gauge 1 × 1" square tube × 60"	2
B	Crossbars	16-gauge 1 × 1" square tube × 25"	2
C	Verticals	16-gauge ½ × ½" square tube × 54"	4
D	Main arch	16-gauge 1 × 1" square tube × 51"*	2
E	Sleeves	⅛ × ½ × ½" angle iron × 2"	2
F	Decorative arches	¼" round rod × varied*	8
G	Mounting stakes	¾" round rod × 36"	2

* Approximate dimensions, cut to fit.

BEND THE MAIN ARCHES

1. Cut a 28"-radius semicircle from ½" plywood. Cut the semicircle in half.
2. Stack the quarter circles and screw them onto a 3 × 3 sheet of plywood. Screw 2 × 4 stops along the bottom.
3. Clamp this bending jig to a stable work surface. Slide an arch piece (D) between the stops so it extends 2" below the bottom of the semicircle. Bend the tube around the curve (see photo, below). Bend the second arch in the same manner.

FINISH THE MAIN ARCHES

1. Remove the quarter circles from the bending jig. From the right angles of the quarter circles, measure in 17½" along one flat side. At this point, draw a line perpendicular to the side. Cut along the line. Repeat with the second quarter circle.

2. Mount these two arcs at the top of a 4 × 8 sheet of plywood, abutting them together on the new cuts to create a template for the arch.
3. Place one arch into the left side of the jig. Mark a vertical line on the arch to match the joint of the two wood arcs (see photo, opposite bottom left). Cut the arch along the line.
4. Place the other arch into the right side of the jig. Mark the centerline and cut.
5. Place both arches back into the jig and weld together at the peak.
6. Cut the sleeves (E) to length. Slide a sleeve 1" into each end of the arch. Slide the top of the panel over the sleeves. Weld the arch to the legs.

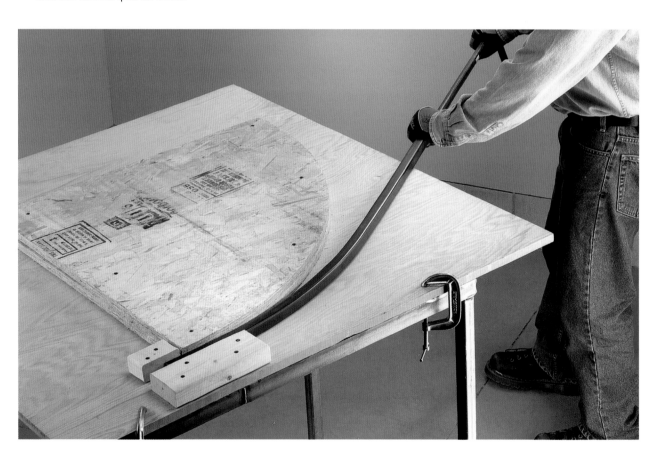

Bend the arch around the jig. Insert a ½" black pipe into the tube end to create more leverage.

MAKE THE DECORATIVE ARCHES

1. Attach a stop block ½" from the base of the arch jig. Bend a decorative arch piece (F) around the jig to make a curve. Continue bending curves as needed.
2. At the 5" mark on the top crossbar, center one end of the arch piece. Align the arch parallel to the main arch. Mark the arch where it crosses the main arch (see photo below, right). Cut on this line. At both ends, tack weld in place.
3. Repeat step 2 at the 10", 15", and 20" marks.
4. Repeat the process of placing, cutting, and welding arches for arches curving in the opposite direction. Finish all welds.
5. To prevent rattling, tack weld the arches to each other where they cross.

FINISH THE ARBOR

1. Wire brush or sandblast the arbor. Finish as desired.
2. Pound the mounting stakes (G) 2' into the ground, 26" apart. Slide the legs over the mounting stakes.
3. Using wire, secure the top or sides of the arbor to a wall or fence.

Place an arch in the second jig and mark where the arc crosses the vertical line between the two half circles.

After bending the decorative arches, mark where they cross the crossbar and main arches.

ARCH & GATE

What a grand entryway this arch and gate provides! Unlike many manufactured archways, this one stands a full eight feet tall, allowing easy passage underneath and plenty of room for trailing foliage. It is proportioned to fit over a 42-inch walkway. The arching pattern is beautiful and fairly easy to create. When paired with the matching vineyard trellis (page 194), this project will really add class to your yard or garden.

MATERIALS

- 16-gauge $^1/_2 \times ^1/_2$" square tube (48$^1/_2$')
- 16-gauge 1 × 1" square tube (70')
- $^1/_8 \times ^1/_2 \times ^1/_2$" angle iron (8")
- $^1/_4$" round rod (48')
- $^3/_4$" round tube (8')
- 2" weldable barrel hinges (2 pair, Decorative Iron #14.1120)
- Gate latch (1, Decorative Iron #14.2012)
- $^1/_2$" plywood (2, 4 × 8) sheets

Side view

Leg arch detail

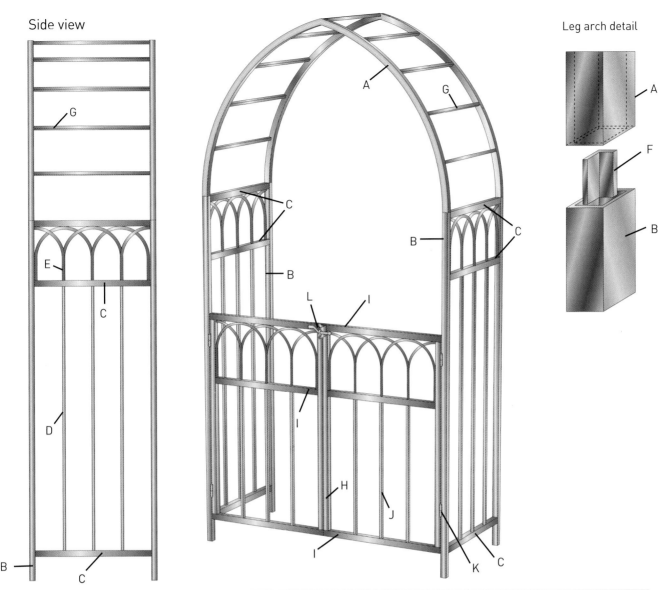

PART	NAME	DIMENSIONS	QUANTITY
A	Main arch	16-gauge 1 × 1" square tube × 51"*	4
B	Legs	16-gauge 1 × 1" square tube × 60"	4
C	Crossbars	16-gauge 1 × 1" square tube × 20"	6
D	Decorative vertical	16-gauge ½ × ½" square tube × 44"	6
E	Decorative arch	¼" round rod × 36"*	16
F	Sleeves	⅛ × ½ × ½" angle iron × 2"	2
G	Arch crossbars	16-gauge ½ × ½" square tube × 20"	9
H	Gate uprights	16-gauge × 1 × 1" square tube × 34"	4
I	Gate crossbars	16-gauge 1 × 1" square tube × 22⅜"	6
J	Gate verticals	16-gauge ½ × ½" square tube × 23"	6
K	Barrel hinges	3"	4
L	Latch		1
M	Mounting stakes	¾" round tube × 24"	4

* Approximate dimensions, cut to fit.

HOW TO BUILD AN ARCH GATE

Before welding, clean all parts thoroughly with denatured alcohol.

MAKE THE MAIN ARCHES

1. Cut a 28"-radius semicircle from ½" plywood. Cut the semicircle in half.
2. Stack the quarter circles and screw them onto a 3 × 3 sheet of plywood. Screw 2 × 4 stops along the bottom.
3. Clamp this bending jig to a stable work surface. Slide an arch piece (A) between the stops, so it extends 4" below the bottom of the semicircle. Bend the tube around the curve (see photo, page 201 top). Bend the remaining arches.

MAKE THE SIDE PANELS

1. Cut the legs (B) and crossbars (C) to length.
2. Lay out two legs and three crossbars on a flat surface. Align the top crossbar flush with the top of the legs and the base of the bottom crossbar 4" from the bottom of the legs (see photo, below). Align the top of the third crossbar 10" down from the top of the legs. Check for square, clamp in place, and tack weld. Repeat to form the second side panel.
3. Before cutting the decorative verticals (D), measure between the bottom and middle crossbars for an exact length. Cut the decorative verticals to length. Mark all three crossbars at 5", 10", and 15".
4. Center the verticals at the marks. Check for square and weld in place. Finish all welds.

Weld the crossbars to the legs.

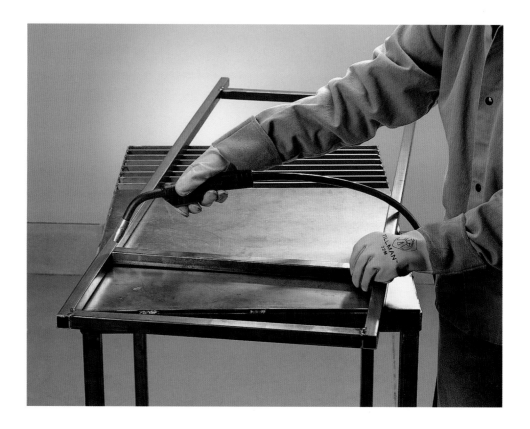

CREATE THE DECORATIVE ARCHES

1. Make a bending jig by attaching a 9"-diameter half circle and 10"-long squares to a 15 × 19" rectangle. Slide a decorative arch piece (E) between the stops and bend it around the curve (see photo, top right).
2. Place two hoops between the top and middle crossbars aligned with the legs and the 10" mark. Mark where the legs cross the middle crossbar, cut, and weld in place.
3. Place a hoop with one leg at the 5" mark and the other at the 15" mark. Mark, cut, and weld (see photo, center right middle).
4. Cut a hoop in half and place the halves starting from the 5" and 15". Weld in place.
5. Repeat steps 2 to 4 for the other side panel.

FINISH THE ARCHES

1. Remove the quarter circles from the bending jig. From the right angles of the quarter circles, measure in 7½" along one flat side. At this point, draw a line perpendicular to the side (see photo, bottom right). Cut along the line. Repeat with the second quarter circle.
2. Mount these two arches at the top of a 4 × 8 sheet of plywood, abutting them together on the new cuts to create a template for the arches. Place stop blocks 1" from the arch.
3. Place the other arch into the right side of the jig. Mark the centerline, and cut.
4. Place both arches back into the jig and weld together at the peak. Repeat steps 3 and 4 for the second arch. Remove the arch jig from the plywood sheet.

ASSEMBLE THE ARCHWAY

1. Cut the sleeves (F) to length. Cut the arch crossbars (G) to length.
2. Mark two parallel lines 45" apart on the sides of the plywood. Attach 2 × 4 clamping blocks along these lines.
3. Clamp the two side panels, lying on their backs, to the jig.
4. Slide sleeves into the legs 1", and weld in place.
5. Slide an arch over the sleeves and weld in place (see photo, page 202). Repeat with the second arch.
6. Weld a crossbar (G) between the two arches at the peak. Weld crossbars spaced 9" apart on the sides of the arches (see photo, page 203, top).

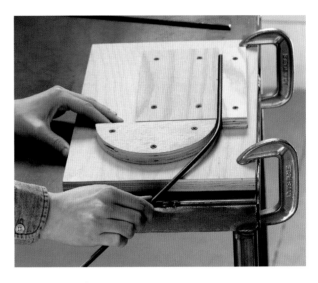

Make a bending jig to create the hoops for the decorative arches.

Align the top of the hoop with the top crossbar and mark where the legs cross the middle crossbar.

After bending the main arches, remove the quarter circles from the jig. Mark in 7½" from one flat edge and cut the form down.

MAKE THE GATE FRAME

1. Cut the gate uprights (H) and crossbars (I) to length.
2. Lay out the two gate frames with the crossbars on the top and bottom of the uprights. Place the third crossbar 10" from the top of the uprights. Check for square, and tack weld together. Check the diagonals and adjust if necessary. Finish the welds.
3. Cut the gate verticals (J) to length. Mark all three crossbars at $5\frac{3}{16}$", $10\frac{3}{16}$", and $15\frac{3}{16}$". Center the verticals at the marks. Check for square, and weld in place (see photo, opposite center).
4. Make and assemble the gate decorative arches using the same method as described under Create the Decorative Arches on page 201.

INSTALL THE GATE

1. Stand the archway upright. Using scrap metal or wood, clamp the archway legs so that the space between the legs is 45" and the panels are parallel.

2. Align the gates flush with a set of legs (the legs and gate are flush front to back) and clamp in place. Allow a slight gap between the gates and legs (the gap is between the legs and gate side to side).
3. Tack weld the hinges (K) in place (see photo, opposite bottom). Remove the clamps holding the gates and check that the gates swing freely. Adjust if necessary and complete the welds. Install the gate latch (L).

FINISH THE ARCHWAY

1. If you plan to finish the archway, sand or wire brush or sandblast it. Apply the finish of your choice.
2. Install the archway by clamping the side panels 45" apart.
3. On the ground where the arch will be installed, mark a 21×46" rectangle. At the corners of the rectangle, pound the mounting stakes 1' into the ground. Lift the archway onto the mounting stakes and remove the clamps.

Clamp the sides to clamping blocks on a 4×8 sheet of plywood. Slide the main arches over the angle iron sleeves in the legs.

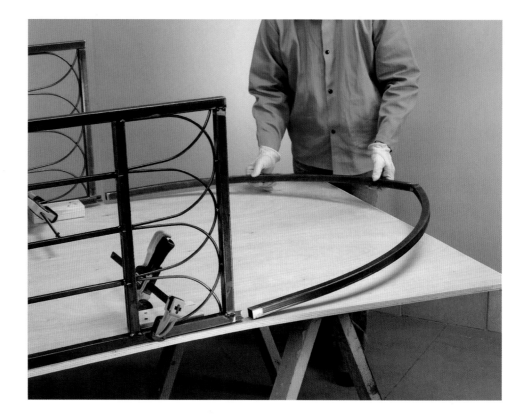

Weld the arch crossbars to the arches.

Center the gate verticals at the marks and weld in place.

Use shims and clamps to hold the gates in place. Weld the base, pin the side of the hinge to the frame, and the top to the gate.

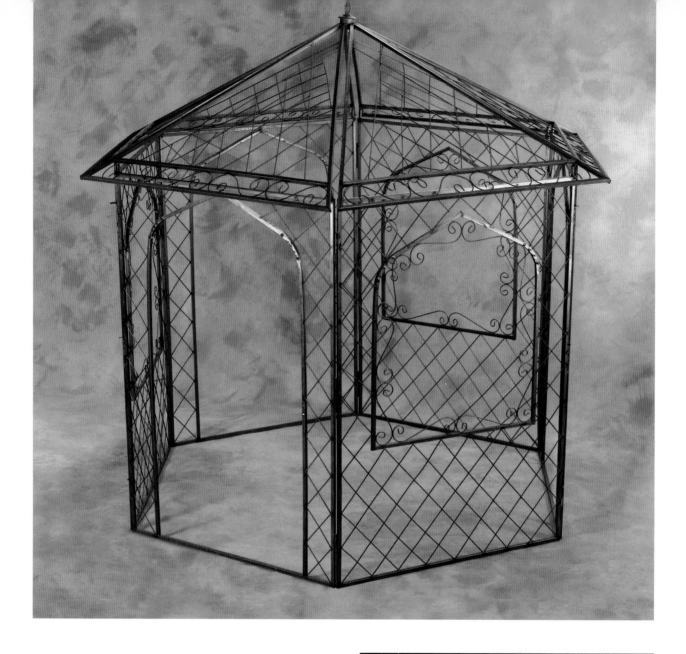

GAZEBO

This moderate-size gazebo is also a special addition to any yard. Big enough for a table and chairs, it is small enough to fit in all but the tiniest of yards. Covered with vines in the summer, it provides delightful shade. In the off-season it supplies a point of visual interest with its scrolling patterns.

Want a different scroll pattern? Simply peruse the online catalogs of the companies listed in the Resources on page 236 for dozens of alternative patterns.

The directions here describe how to build door, window, and roof panels. The material quantities are for a gazebo with three door panels and three window panels. You can choose to construct your gazebo with any combination, just remember to adjust your quantities list accordingly.

MATERIALS

- 4" finial (1, Decorative Iron #126)
- 1 × 1" square tube (344')
- 3/16" round rod (1,000')
- 1/2" round tube (6')
- 3/8" round rod (10½')
- 1/4" round rod (3')
- 4½" drip trays (2)
- 2" round tube (2")
- 1/2 × 6" carriage bolt, nut, and washer (1)
- 3/4" round rod or rebar (9')

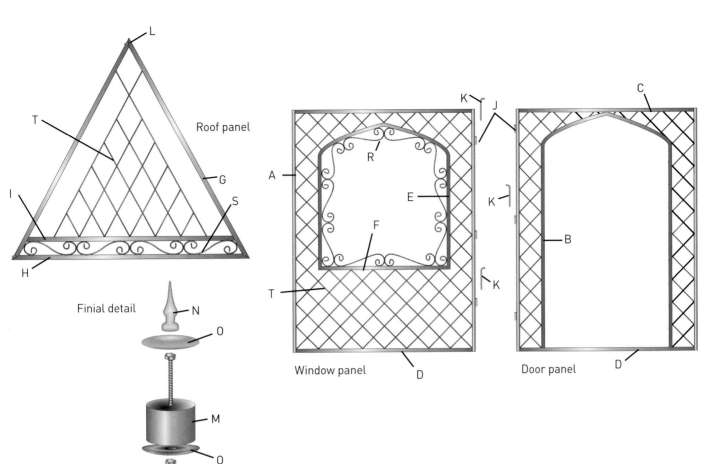

Roof panel

Finial detail

Window panel

Door panel

PART	NAME	DIMENSIONS	QUANTITY
A	Legs	16-gauge 1 × 1" square tube × 80"	12
B	Door frames	16-gauge 1 × 1" square tube × 96"	6
C	Top crossbars	16-gauge 1 × 1" square tube × 54"	6
D	Bottom crossbars	16-gauge 1 × 1" square tube × 52"	6
E	Window frames	16-gauge 1 × 1" square tube × 40"	6
F	Window crossbars	16-gauge 1 × 1" square tube × 30"*	6
G	Roof sides	16-gauge 1 × 1" square tube × 66½"	12
H	Roof bases	16-gauge 1 × 1" square tube × 61"	6
I	Roof crossbars	16-gauge 1 × 1" square tube × 61"*	6
J	Hinge barrels	16-gauge ½" round tube × 3"	24
K	Hinge pins	⅜" round rod × 7½"	12
L	Finial pegs	¼" round rod × 6"	6
M	Finial sleeve	16-gauge 2" round tube × 2"	1
N	Finial	4"	1
O	Drip trays	4½"	2
P	Roof pegs	⅜" round rod × 3"	12
Q	Peg sleeves	16-gauge ½" round tube × 3"	12
R	Window scrolls	3⁄16" round rod × 24"	30
S	Roof scrolls	3⁄16" round rod × 36"	24
T	Diagonals	3⁄16" round rod × varied*	624

* Cut to fit.

HOW TO BUILD A GAZEBO

Before welding, thoroughly clean all parts with denatured alcohol.

MAKE THE DOOR PANELS

1. Cut two legs (A), two door frames (B), a top crossbar (C), and bottom crossbar (D) to length. Miter one end of each leg at 45° and both ends of the top crossbar at 45°.
2. Lay out the legs, top crossbar, and bottom crossbar to form a rectangle. The bottom crossbar fits between the legs.
3. Check the rectangle for square, and tack weld all corners.
4. Mark the door frames at 24". Using a conduit bender, bend the frames at the mark to 50° (see photo, below).
5. Place the door frames inside the assembled panel frame. Align the door frames 8" in from the panel sides. Mark vertical lines on the door frames where they overlap at the peak. Cut on the marks, and tack weld the door frame arch together.
6. Align the door frame arch inside the panel frame with the point of the arch touching the top crossbar. Mark the legs where they cross the bottom crossbar. Cut the legs to length. Tack weld the door frame to the top and bottom crossbars.
7. Repeat steps 1 through 6 to make the desired number of door panels. Complete the welds.

MAKE THE WINDOW PANELS

1. Cut two window frames (E) and a window crossbar (F) to length.
2. Mark the window frames at 24". Using a conduit bender, bend the frames at the mark to 50° (see photo, right).
3. Lay out the window frames with the window crossbar between two legs (A). Mark the vertical line at the peak where the frames cross each other. Cut the frames on the line, and tack weld together.
4. Trim the legs to length, if necessary. Tack weld the window crossbar in place. Put the window frame aside for later installation.
5. Repeat steps 1 through 4 to make the desired number of window panels.

MAKE THE ROOF PANELS

1. Cut two roof sides (G) to length. Miter one end of each side at 29°. Align the mitered edges, and tack weld.
2. Cut a roof base (H) to size. Evenly align it underneath the sides. Mark where the sides cross the base. Cut on this line. Tack weld the base to the sides.
3. Cut a roof crossbar (I) to size. Place the crossbar 5" up from the roof base and mark the angles where the roof sides cross the crossbar (see photo, opposite right).
4. Repeat steps 1 to 3 to make the remaining five roof panels. Complete the welds.

ASSEMBLE THE GAZEBO

Assembling the gazebo before all the decorations are added makes it lighter and easier to maneuver. In addition, if a panel is slightly skewed, it is easier to make corrections without having to rearrange decorative elements. Because the hinges will be perfectly aligned only for one arrangement, make sure you assemble the panels in the order desired.

1. Cut the hinge barrels (J) and pins (K) to length. With a bench vise, bend a right angle ½" in from one end of each pin.
2. Slide a pin into two barrels. Place two panels on the floor or upright against a wall. Place the barrels against the joint between the panels at 20" from the top. Tack weld one barrel to one panel and the other barrel to the second panel (see photo, opposite left).
3. Tack weld another hinge to the panels at about 20" from the bottom.
4. Repeat steps 2 and 3 with the two other sets of two panels.
5. With the panels standing, arrange the three sets of two panels to form a hexagon. The angle between each panel pair should be 120°.
6. Install hinges at the remaining three panel joints.

Create the window and door arches by bending the tubing with a large conduit bender.

ASSEMBLE THE FINIAL

1. Cut the finial pegs (L) to length. Bend the finial pegs in half to 110°. Support the roof sections on a cinder block or stool, and tack weld a finial peg to the point of each roof section (see photo, page 208, left).
2. Cut the finial sleeve (M) to length.
3. Weld the finial (N) to the center of one drip tray (O) on the convex side.
4. Grind down the head of a $\frac{1}{2} \times 6"$ carriage bolt until it matches the concave curve of the drip tray. Weld the carriage bolt to the middle of the concave side of the drip tray from step 3 (see photo, page 208, right).
5. Center the finial sleeve over the bolt and weld in place. Drill a $\frac{9}{16}"$ hole through the center of the second drip tray.
6. With the peaks of the roof panels resting on a cinder block or low stool, slide the finial assembly over the finial pegs. Bend the finial pegs if necessary to fit inside the finial sleeve.

ASSEMBLE THE ROOF

Assembling the roof is easiest with an assistant or two.

1. Cut the roof pegs (P) and peg sleeves (Q) to length.
2. Weld two sleeves to the tops of each side panel $\frac{3}{4}"$ in from the sides of the legs (see photo, page 209, left).
3. Measure the distance between a pair of sleeves. Use this measurement to mark the peg locations on a roof panel. Grind the pegs to 60° on one end, and weld the ground end to the roof panel.
4. With the side panels assembled, place a roof section on top of a wall panel, sliding the pegs into the sleeves. Have an assistant hold up the center of the roof, or create a wood support.
5. Place the remaining roof sections. Slide the finial over the finial pegs, slide the drip cup over the bolt, and install the washer and nut.
6. Adjust roof and finial pegs as needed.
7. Number or otherwise mark the side and roof panels for reassembly.
8. Remove the finial and take the roof panels down. Disassemble the panels by pounding out the hinge pins.

Mark the roof crossbar for cutting.

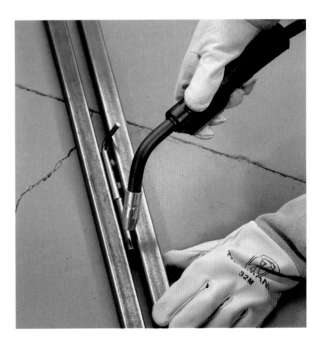

Insert a hinge pin into two hinge barrels. Place the hinge between two side panels, and weld one barrel to one panel and the other barrel to the adjoining panel.

Support roof sections on a cinder block. Weld the finial pegs to the points of the panels.

Create the finial assembly by welding the finial to the drip tray, then welding the bolt to the drip tray.

MAKE THE WINDOW SCROLLS

1. Cut the scroll blanks (R) to length.
2. Cut a 3/16" slot in a 2" pipe. Place the rod in both slots, then bend the rod around the outside of the pipe one and a half times (see photo, opposite right).
3. Place the other end of the rod in the slot and bend one and a half turns around the outside of the pipe in the opposite direction.
4. Slightly open the bends to create a pleasing scroll.
5. Repeat to form the remaining window scrolls.
6. Weld the scrolls into the window frames as pictured.

MAKE THE ROOF SCROLLS

1. Cut the scroll blanks (S) to length.
2. Cut a 3/16" slot in a 3" pipe. Place the rod through both slots, then bend the rod around the outside of the pipe one and a half times.
3. Place the other end of the rod in the slot and bend one and a half turns around the outside of the pipe in the opposite direction.

4. Slightly open or close the bends to make the scroll fit between the roof base and roof crossbar.
5. Repeat for the other roof scrolls.
6. Weld the scrolls between the roof base and roof crossbar.

ADD DECORATION TO THE ROOF

You can make the roof a simple diamond pattern.

1. To make the diagonals (T), mark the roof sides every 6" down from the peak and 6" down from the roof crossbar. Lay out two rods from the peak marks to the opposite first two marks on the roof crossbar to match the roof panel diagram (page 205). Cut the rods to fit, and weld in place.
2. Mark the diagonals from step 1 every 6".
3. Make the internal diagonals by cutting pieces to fit between the marks on the right side and base. Weld in place.
4. Finish the internal diagonals by cutting pieces to fit between the marks on the left side and base. Weld in place.

Weld the roof sleeves to the tops of the side panels.

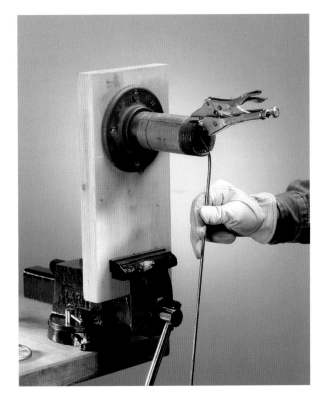
Cut a ³⁄₁₆" notch in a 2" pipe to create the bending jig for the scrolls. Slide the rod into the notch, then bend the rod around the pipe.

ADD DIAGONAL FILLER TO SIDE PANELS

Diagonal rods can fill the space below and above the windows, or you can add scrolls and hearts.

1. On the door panels, mark the legs, door frames, and top crossbars every 8".
2. Cut diagonals to fit between the marks.
3. Weld the descending left to right diagonals in place, then weld the ascending left to right diagonals in place.
4. On the window panels, mark the legs, top and bottom crossbars, window frames, and window crossbars every 8". Align the window frame between the legs, 8" in from each side with the 8" marks aligned. The bottom of the window should be 32" from the bottom crossbar.
5. Cut diagonals to fit between the marks on the window frame and legs.
6. Cut diagonals to fit between the marks on the crossbars and the window frames and window crossbars.
7. Weld the descending left to right diagonals in place, then weld the ascending left to right diagonals in place.

FINISH THE GAZEBO

1. Wire brush or sand blast the gazebo. Finish as desired.
   Note: Take care to track the panel numbers so you can reassemble the gazebo.
2. Reassemble the gazebo.
3. For at least three corners of the gazebo, pound 3' lengths of ¾" round bar or 18" rebar into the ground, and slide the leg ends over the bars.

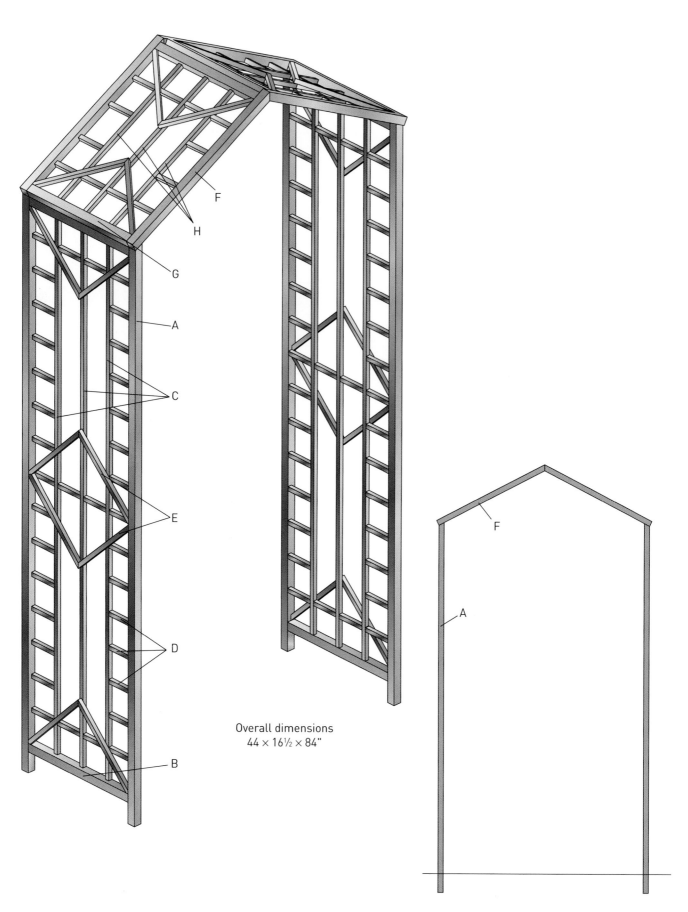

Overall dimensions
44 × 16½ × 84"

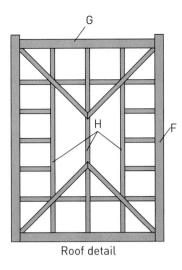

Roof detail

ARBOR

This distinctive arbor will add a touch of class to any garden. Modeled after the prairie style, its clean lines are straightforward to cut and weld, and they provide plenty of climbing support for a variety of vines. The ½" inserts are placed flush with the back of the 1" sides and ends, eliminating the need for difficult centering and giving an increased sense of depth. The arbor can be made from the 6' steel lengths available at home improvement centers, but because of the amount of steel used, a trip to a steel supplier might be worthwhile, especially if you have a way to transport 20' lengths. The ½" square tube bows quite easily, so if you do buy 20' lengths, make sure they are well supported in transit. Otherwise, they will bow and not be suitable for the vertical inserts.

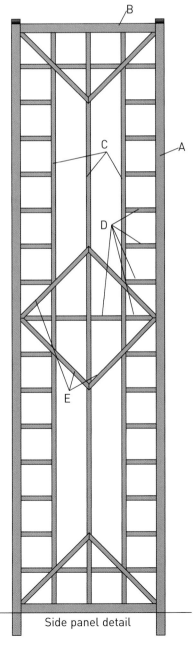

Side panel detail

PART	NAME	DIMENSIONS	QUANTITY
A	Panel sides	16-gauge 1 × 1" square tube × 72"	4
B	Panel ends	16-gauge 1 × 1" square tube × 15½"	4
C	Panel vertical inserts	16-gauge ½ × ½" square tube × 63½"	6
D	Horizontal inserts	16-gauge ½ × ½" square tube × 3½"	100
E	Diagonal inserts	16-gauge ½ × ½" square tube × 12"*	24
F	Roof ends	16-gauge 1 × 1" square tube × 15½"	4
G	Roof sides	16-gauge 1 × 1" square tube × 25½"	4
H	Roof vertical inserts	16-gauge ½ × ½" square tube × 23½"	6

*Approximate dimension, cut to fit.

MATERIALS

- 16-gauge 1 × 1" square tube (44')
- 16-gauge ½ × ½" square tube (100')
- Wood scraps

Check the panel for square by measuring the diagonals. If the measurements are equal, the assembly is square.

Weld the diagonal inserts to the panel framework and to the center vertical insert.

HOW TO BUILD AN ARBOR
ASSEMBLE THE PANELS

Set up the project on a sheet of ¾" plywood placed on sawhorses. Make sure the plywood is not bowed or it will cause misalignment of the project pieces. Working on a raised surface is easier than working on the floor, and you can clamp the metal to the plywood. Remember to clamp your work cable to the workpiece. If you want the arbor wider or narrower than 44", you must adjust the miter angles for the roof and side panels.

1. Cut the panel sides (A) to size, mitering one end at 30°. The mitered end is the top. Cut the panel ends (B), vertical inserts (C), and horizontal inserts (D) to size.

2. Place a panel end between the top of two panel sides, keeping the outside edge of the panel end flush with the short ends of the mitered panel sides. Clamp the pieces in place.

3. Position three vertical inserts between the two panel sides. (This is just to hold the bottom panel end in place. The exact location of the inserts is not important at this time.)

4. Place the bottom panel end between the panel sides, keeping it snug against the vertical inserts. Clamp the bottom panel end in place.

5. Remove the vertical inserts. Use a carpenter's square to check the panel sides and ends for square. Tack weld each corner. Check for square again by measuring the two diagonals (see photo, left). If the measurements are equal, the panel is square. If not, adjust the workpieces until the diagonals are the same.

6. Turn the panel over and complete the welds at the four corners. Grind the tack welds flat so the panel lies flat on the work surface.

ATTACH THE VERTICAL & HORIZONTAL INSERTS

1. Replace the vertical inserts between the panel sides. Place several horizontal inserts between the vertical inserts and the panel sides to ensure the correct spacing of the vertical inserts.

2. Weld the vertical inserts to the panel top and bottom.

3. Place the horizontal inserts into the vertical framework every 3½". (Use horizontal inserts as spacers.)

4. Starting at the top of the panel, use a combination square to align the first row of horizontal inserts. Once they are aligned, weld them in place.

5. Continue aligning and welding the inserts one row at a time. If the vertical inserts are bowed side to side, use a clamp to hold them against the inserts. If they bow upward, weigh them down with a sandbag or other weight.

INSERT THE DIAGONAL INSERTS

1. Cut four diagonal inserts (E) to size, mitering one end at 45°.

2. Place two diagonal inserts against the bottom edge of the top panel. The mitered ends of the inserts should butt together over the middle vertical insert. Mark the other end of the diagonal inserts where they cross the top panel so they will fit in the corner.

3. Make the angled cuts on the diagonal inserts.

4. Grind down the welds that will be underneath the diagonal inserts so they will lie flat. Weld the diagonal inserts in place (see photo, opposite bottom).

5. Repeat steps 2 to 4 to weld the diagonal inserts to the bottom of the panel.

6. Cut four more diagonal inserts, mitering both ends at 45°.

7. Center these four inserts over the eighth row of horizontal inserts to form a diamond shape. Weld the inserts in place.

8. Repeat this entire process to construct a second panel.

Make clamping blocks out of wood scraps. Clamp the roof panels to the blocks. Place two or three tack welds along the joint between the panel ends.

BUILD THE ROOF PANELS

1. Construct two roof panels following the same procedure used for the side panels, except with no central diamond insert.

2. Stand the roof panels on edge and join the mitered ends to form the 120° angle for the roof peak.

3. Clamp the roof panels in place. (You may need to attach temporary wooden clamping points to the work surface by fastening 1 × 2 or 2 × 4 scraps to match the layout. Clamp the panels to the scraps.) Place two or three tack welds along the joint between the panel ends (see photo, page 213).

FASTEN THE ROOF TO THE SIDES

1. Place a side panel on edge, and set it against one side of the roof panel. Line them up so the mitered edge of the side panel is making contact with the bottom edge of the roof panel. Clamp the panels in place.

2. Tack weld the panels at two points along the joint (see photo, below).

3. Repeat steps 1 and 2 for the other side panel.

4. Measure the distance between the tops and the bottoms of the side panels to ensure the panels are an equal distance apart at both ends.

5. Check the panels for square by measuring the diagonals. Make any necessary adjustments, and make sure the roof assembly is still fitted against the side panels.

6. Weld all the mitered roof peak joints and the mitered joints between the side panels and the roof panels.

FINISHING TOUCHES

To prevent putting stress on the welds, and to maintain the arbor's shape, clamp wooden crosspieces between the panel sides before moving it.

1. The easiest finish for the arbor is to allow it to gently rust over the years. You may want to place plastic end caps over the exposed tube ends on the roof, or you can weld on small caps.

2. The arbor can sit on the ground or can be mounted in concrete footings. If you live in a windy location, you may want to drive four lengths of rebar into the ground and slip the arbor legs over them.

Place tack welds along the joint between the roof and side panels.

PATIO BENCH

Finding time to relax just became easier. Create a private area to place this custom patio bench to enjoy those evenings listening to the sounds of nature or enjoy the stars.

MATERIALS

- 1 × 1" 11-gauge HRS square tubing (two 24′ lengths)
- 19 × 15" 10-gauge HRS sheet
- 1 × 1" 11-gauge HRS angle iron (40" length)
- ³⁄₁₆ × 1¼" HRS flat bar (eleven 10′ lengths)

PART	DIMENSIONS	STOCK	QUANTITY
A	35¹³⁄₁₆"	1 × 1" 11-gauge HRS square tubing	2
B	16"	1 × 1" 11-gauge HRS square tubing	3
C	28"	1 × 1" 11-gauge HRS square tubing	3
D	22"	1 × 1" 11-gauge HRS square tubing	3
E	20"	1 × 1" 11-gauge HRS square tubing	6
F	16½"	1 × 1" 11-gauge HRS square tubing	4
G	4¾ × 18"	10-gauge HRS sheet	3
H	1"	1 × 1" 11-gauge HRS angle iron	30
I	38"	³⁄₁₆ × 1¼" HRS flat bar	33

HOW TO BUILD A PATIO BENCH

CUT & ASSEMBLE THE FRAME

Cut parts A and B. Position part B between the two pieces of part A, centered along the lengths. Measure and check for squareness, and then weld all joints of this frame (see photo, below).

1. Cut three lengths of part C. Cut one end of each piece at an 80° angle, removing approximately ¼" of material. Assemble and tack on one piece of part C at each end of parts A. Make sure the angled end is attached along the frame to allow for a 10° slope to the back.
2. Cut three lengths of part D. Cut six lengths of part E, mitering each end at a 10° angle. Make sure the angles are parallel to one another. Cut four pieces of part F.
3. Align and tack weld one leg to each corner of the frame, and center two additional legs along the length of part A (see photo, opposite top).
4. Align and tack weld braces F and D (see photo, opposite bottom).

Measure and check for squareness, and then weld all joints of this frame.

Align and tack weld one leg to each corner of the frame.

Tack weld brace F to brace D.

Layout and cut the seat sides, as shown.

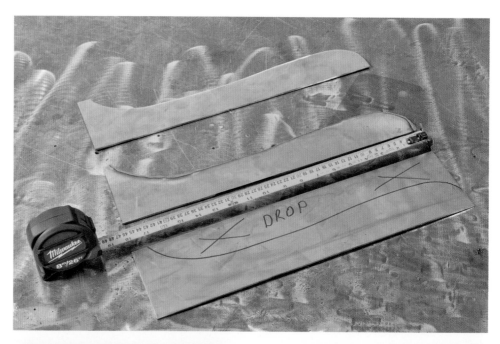

Attach the angle iron to the seat sides.

LAY OUT & ASSEMBLE THE SEAT

1. Lay out and cut the seat sides, Part G, as shown (see photo above, top).
2. Cut the 30 pieces of part H. Attach the two seat sides (part G) (see photo above, bottom).
3. Weld the seat sides to the frame (see photo opposite, top).
4. Cut the third seat plate to fit along the center brace as shown.

ATTACH SEAT & BACK

Cut the 33 seat backpieces (I). Attach them, spaced 2" on center; allow for a ⅛" space between the slats (see photo opposite, bottom).

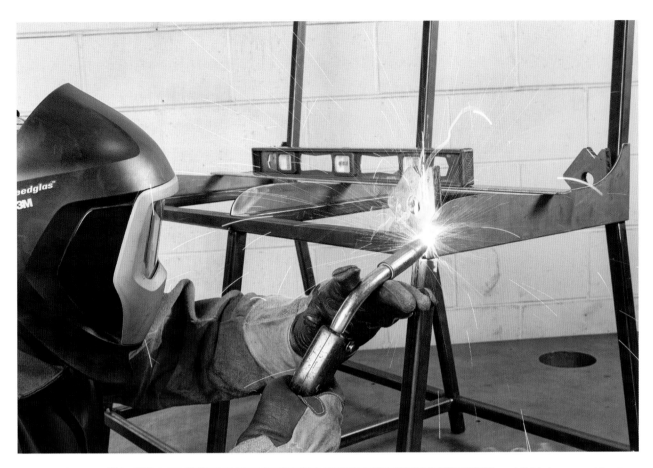

Weld the seat sides to the frame.

Attach the seat and back material.

GARDEN TUBE WATERFALL

Fabricate a work of art that will also provide a calming effect for any room or living space. The sound of the water gently moving through the waterfall and cascading over the rocks will help provide a relaxing atmosphere after a day of work.

Work with your local home center to locate the appropriate pump and tube. Ask for help in the garden center with water garden supplies.

Before welding, clean all parts to remove mill scale and any oil residue. Wipe clean with denatured alcohol, then grind any remaining mill scale ½" wide through any areas to be welded.

MATERIALS

- 2 × 2" 11-gauge HRS square tubing (18', 3" length)
- 14-gauge CRS sheet (15½ × 36")
- Small water pump
- Decorative gravel or rocks
- Clear plastic tubing to fit pump (60")

PART	DIMENSIONS	STOCK	QUANTITY
A	48"	2 × 2" 11-gauge HRS square tubing	2
B	36"	2 × 2" 11-gauge HRS square tubing	2
C	24"	2 × 2" 11-gauge HRS square tubing	2
D	3 × 6"	14-gauge CRS sheet	2
E	3 × 36"	14-gauge CRS sheet	2
F	6 × 36"	14-gauge CRS sheet	1

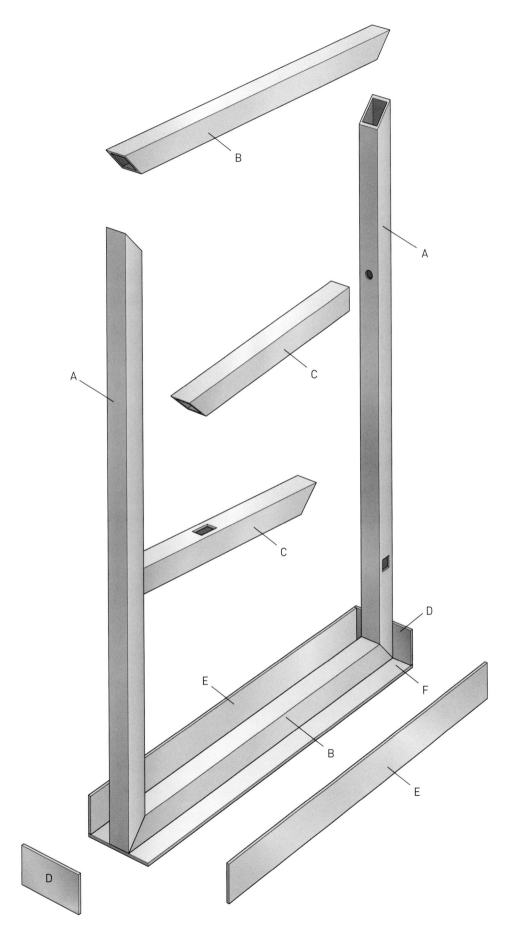

HOW TO BUILD A GARDEN TUBE WATERFALL

FABRICATING THE FRAME

1. Cut two pieces of part A and two pieces of part B, cutting each end to a 45° angle.

2. On one piece of part A, lay out and cut a 1 × 1" square hole, centered 6" from the lower side of the 45° bevel, (see photo below, top). On this same part, cut or drill a ¾" hole on center, 12" up from the short point of the 45° bevel.

3. Tack the frame together as shown. Measure to check for squareness (see photo below, bottom), then weld all joints. Remember, the lower weld joints must be watertight.

Cut a 1 × 1" hole in the center of part A.

Check for square.

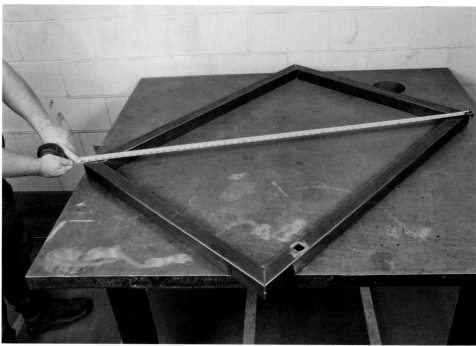

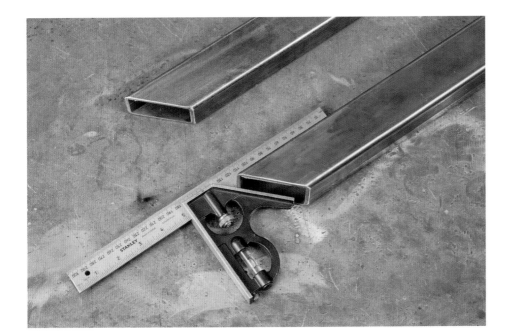

Cut one end of square tubing to 45° angle.

Cut a rectangular hole on center of part C.

ADDING THE CROSS TUBE

1. Cut two pieces of part C, cutting one end to a 45° angle, the other end to a 5° angle (see photo above, top).
2. On one end of part C, lay out and cut a 1¼ × 2" rectangular hole on center, 9" from the upper side of the 5° bevel (see photo above, bottom).
3. Center and weld the cross tubes.

FABRICATE THE RESERVOIR

1. Cut parts D, E, and F. Center and tack part D, one on each end parts A.
2. Align and tack part F along the underside of the frame and part D.
3. Align and tack part E, one on each side of part D, to form the reservoir.

4. Weld all seams in small increments, 1" to 2" apart. These seams must be watertight (see photo below, top).
5. Add pump gravel, and feed plastic tube through the vertical upright, as shown (see photo below, bottom).

Weld the seams of the reservoir.

Feed the plastic tube through the vertical upright.

APPENDIX: ADDITIONAL PLANS

SCROLL DESK LAMP

Overall dimensions
8 × 15"

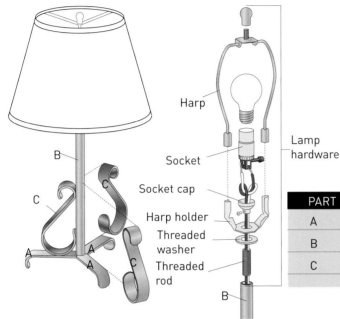

Harp

Lamp hardware

Socket

Socket cap

Harp holder

Threaded washer

Threaded rod

MATERIALS

- $1/8 × 3/4$" flat bar (15")
- $1/2$" round tube (9")
- $1/8 × 1/2$" flat bar (36")
- $1/2$", 1", and 2" pipe
- Lamp hardware and shade
- $3/8$" hollow threaded rod (1")

PART	NAME	DIMENSIONS	QUANTITY
A	Bases	$1/8 × 3/4$" flat bar × 5"	3
B	Shaft	$1/2$" round tube × 9"	1
C	Scrolls	$1/8 × 1/2$" flat bar × 12"	3

TREE-SHAPED CANDLEHOLDER

Overall dimensions
16 × 8 × 20"

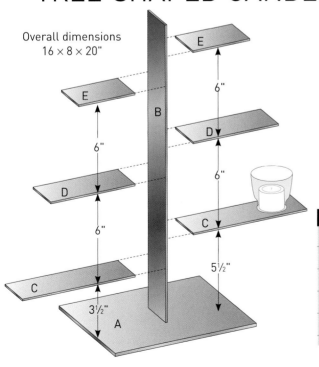

MATERIALS

- $3/16$" sheet (8 × 10")
- $3/16 × 2$" flat bar (5')
- Votive holders and candles
- Silicone adhesive

PART	NAME	DIMENSIONS	QUANTITY
A	Base	$3/16$" sheet × 8 × 10"	1
B	Trunk	$3/16 × 2$" flat bar × 20"	1
C	Bottom branches	$3/16 × 2$" flat bar × 8"	2
D	Middle branches	$3/16 × 2$" flat bar × 6"	2
E	Top branches	$3/16 × 2$" flat bar × 4"	2

FIREPLACE CANDLEHOLDER

Overall dimensions
16 × 11 × 6"

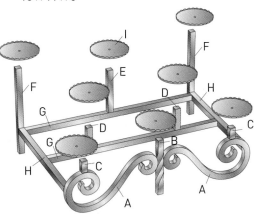

MATERIALS

- 4 × 8" scrolls (2)
- ½" twisted square rod (5")
- ½" square rod (7')
- 3⅜" bobeches (8)
- Candles (8)

PART	NAME	DIMENSIONS	QUANTITY
A	Scrolls	4 × 8"	2
B	Center post	½" twisted square rod × 5"	1
C	Front extenders	½" square rod × 1"	2
D	Middle extenders	½" square rod × 2"	2
E	Rear extender	½" square rod × 3"	1
F	Rear posts	½" square rod × 6"	2
G	Crosspieces	½" square rod × 16"*	2
H	Side pieces	½" square rod × 10"	2
I	Bobeches	3⅜"	8

*Approximate dimensions, cut to fit.

WALL-MOUNTED CANDLEHOLDER

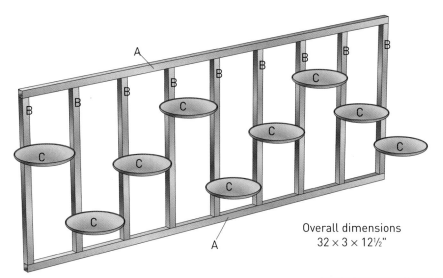

Overall dimensions
32 × 3 × 12½"

MATERIALS

- ½" square bar (14')
- 2¼" bobeche (9)
- Candles (9)

PART	NAME	DIMENSIONS	QUANTITY
A	Horizontal bars	½" square bar × 32"	2
B	Vertical bars	½" square bar × 11½"	9
C	Bobeches	2¼"	9

CHANDELIER

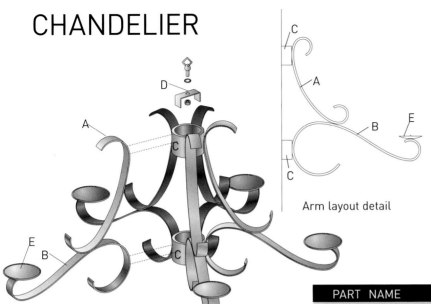

Arm layout detail

MATERIALS

- ⅛ × ¾" flat bar (17')
- 10", 6", 2", and 1" bending forms
- 2¼" round tube (4")
- 2½" bobeches (5)
- 2" threaded nipple
- Lock washers (2)
- Threaded brass washers (3)
- Finial
- Decorative chain
- Candles (5)

PART	NAME	DIMENSIONS	QUANTITY
A	C scrolls	⅛ × ¾" flat bar × 16"	5
B	S scrolls	⅛ × ¾" flat bar × 24"	5
C	Couplers	2¼" round tube × 2"	2
D	Insert	⅛ × ¾" flat bar × 4"	1
E	Bobeches	2½"	5

CORNER COAT RACK

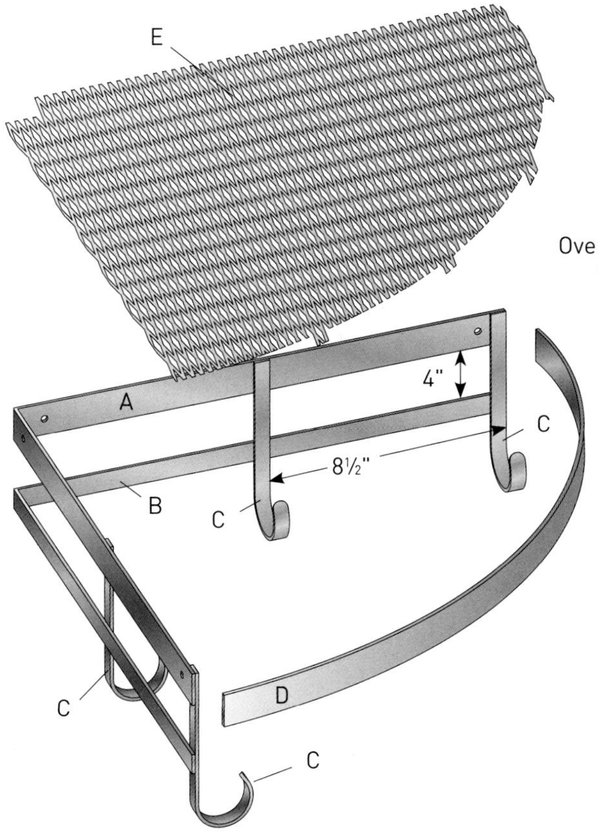

Overall dimensions
18 × 18 × 8"

MATERIALS

- ⅛ × 1¼" flat bar (7')
- ⅛ × ¾" flat bar (7')
- Expanded sheet metal (2 × 2')
- 2" pipe
- Hanging hardware

PART	NAME	DIMENSIONS	QUANTITY
A	Rack top	⅛ × 1¼" flat bar × 36"	1
B	Rack bottom	⅛ × ¾" flat bar × 36"	1
C	Hooks	⅛ × ¾" flat bar × 12"	4
D	Rack front	⅛ × 1¼" flat bar × 38"*	1
E	Shelf	Expanded sheet metal	1

*Approximate dimension, cut to fit.

BAKER'S SHELVES

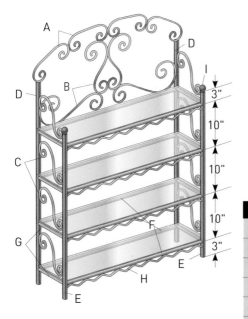

PART	NAME	DIMENSIONS	QUANTITY
A	C scrolls	¼" round bar × 26½"	4
B	Top S scrolls	¼" round bar × 35"	4
C	Side S scrolls	¼" round bar × 24½"	8
D	Back legs	¾" round tube × 44"	2
E	Front legs	¾" round tube × 36"	2
F	Shelf supports, front & back	½" round tube × 34½"	8
G	Shelf supports, sides	½" round tube × 8½"	8
H	Zigzag trim	¼" round bar × 48"	4
I	Finials	1" brass or wood ball	2

TAPERED-LEG TABLE

Square = 1"

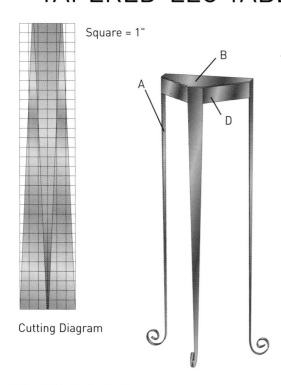

Cutting Diagram

PART	NAME	DIMENSIONS	QUANTITY
A	Legs	⅛" sheet metal × 3 × 30"	3
B	Top	⅛" sheet metal × 12 × 12" right triangle	1
C	Front skirt	⅛ × 2" flat bar × 12"*	1
D	Side skirts	⅛ × 2" flat bar × 7"*	2

* Approximate dimensions, cut to fit.

ONE PIECE PLANT STAND

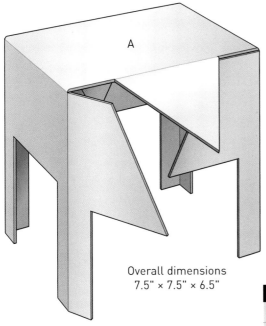

Overall dimensions
7.5" × 7.5" × 6.5"

MATERIALS

- CRS sheet 14 ga. 20.5" × 14"

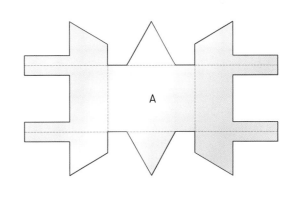

PART	NAME	DIMENSIONS	QUANTITY
A	CRS Sheet 14 ga.	20.5" × 14"	1

COAT TREE

MATERIALS

- 6' Sch 40 1½" pipe
- 30" HRS Bar ⅜" × 1¼"
- 7" HRS Bar ³⁄₁₆" × 1¼" (3)
- 12" HRS Solid ½" (3)
- 24" HRS Solid ½" (3)
- 6" HRS Solid ½" (3)

PART	NAME	DIMENSIONS	QUANTITY
A	Sch 40 Pipe	6' × 1 ½"	1
B	HRS Bar	30 × ⅜ × 1 ¼"	3
C	HRS Bar	7 × ³⁄₁₆ × 1 ¼"	3
D	HRS Solid	12 × ½"	3
E	HRS Solid	24 × ½"	3
F	HRS Solid	6 × ½"	3

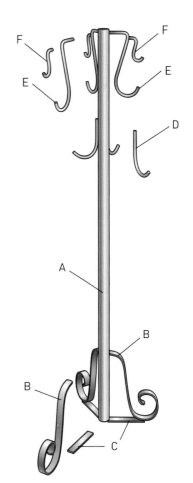

COAT RACK

MATERIALS

- 16-gauge $\frac{1}{2} \times \frac{1}{2}$" square tube (58')
- 16-gauge 1×1" square tube (50')
- 4" O.D. square tube rings (20, Triple S Steel #SR 4)
- $\frac{1}{8} \times 1$" round tube (39")
- $\frac{1}{8} \times \frac{1}{2}$" flat bar (39')
- $1\frac{1}{2}$" spheres (8, Triple S Steel SF116F4)

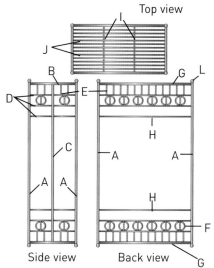

PART	NAME	DIMENSIONS	QUANTITY
A	Legs	16-gauge 1×1" square tube × 68"	4
B	Side crossbars	16-gauge 1×1" square tube × 19"	4
C	Center supports	16-gauge 1×1" square tube × 66"	2
D	Side horizontals	16-gauge $\frac{1}{2} \times \frac{1}{2}$" square tube × 9"	24
E	Verticals	16-gauge $\frac{1}{2} \times \frac{1}{2}$" square tube × 4"	60
F	Circles	4" O.D. square tube rings	20
G	Crossbars	16-gauge 1×1" square tube × 39"	3
H	Back horizontals	16-gauge $\frac{1}{2} \times \frac{1}{2}$" square tube × 39"	6
I	Shelf supports	16-gauge $\frac{1}{2} \times \frac{1}{2}$" square tube × 19"	2
J	Shelf rods	$\frac{1}{8} \times \frac{1}{2}$" flat bar × 39"	12
K	Hanging rod	$\frac{1}{8} \times 1$" round tube × 39"	1
L	Spheres	$1\frac{1}{2}$"	8

WINDOW BOX

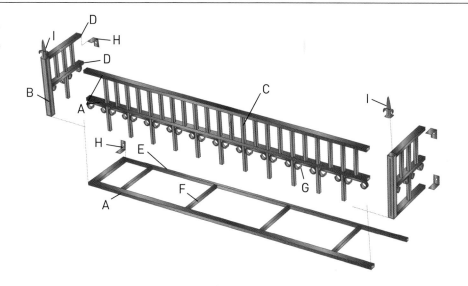

HANGING CHANDELIER

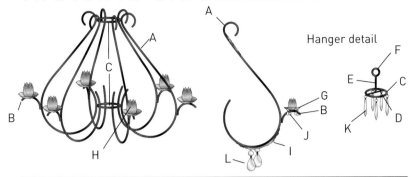

Hanger detail

PART	NAME	DIMENSIONS	QUANTITY
A	Scrolls	³⁄₁₆" round rod × 36"	6
B	Candle arms	³⁄₁₆" round rod × 3"*	6
C	Rings	³⁄₁₆" round rod × 9½"*	2
D	Crossbar	³⁄₁₆" round rod × 3"	1
E	Hanger	³⁄₁₆" round rod × 6"	1
F	Hanging ring	³⁄₁₆" round rod × 4"*	1
G	Bobeches (drip trays)	2³⁄₈"	6
H	Candle cups	1¼ × 1¼"	6
I	Acanthus leaves	1¼ × 8½"	6
J	Teardrop crystals	1"	24
K	Hangdrop crystal prisms	3"	5
L	Teardrop prisms	3"	12

* Approximate dimensions, cut to fit.

MATERIALS

- ³⁄₁₆" round rod (24')
- 2³⁄₈" bobeches (6, Architectural Iron Designs #79/6)
- Candle cups (6, Architectural Iron Designs #78/4)
- 8½" acanthuses (6, Architectural Iron Designs #11942 8½")
- 1" teardrop crystals (24, Chandelierparts)
- 3" hangdrop crystal prisms (5, Chandelierparts)
- 3" teardrop prisms (12, Chandelierparts)

MATERIALS

- 1 × ½ × ⅛" channel (18')
- 1" square tube (20")
- ½" square tube (16')
- ½" square bar (8')
- ⅛ × ½" flat bar (14½')
- ⅛ × 1" flat bar (8')
- Finials (2, Architectural Iron #83/4)

PART	NAME	DIMENSIONS	QUANTITY
A	Front crossbars	1 × ½ × ⅛" channel × 52½"	3
B	End posts	1" 16-gauge square tube × 10"	2
C	Uprights	½" 16-gauge square tube × 4¼"	42
D	Side crossbars	1 × ½ × ⅛" channel × 8⅞"	6
E	Base	½" square bar × 54½"	1
F	Base crossbars	½" square bar × 9¾"	4
G	C scrolls	⅛ × ½" flat bar × 10½"	16
H	Mounting brackets	⅛ × 1" flat bar × 2"	4
I	Finials		2

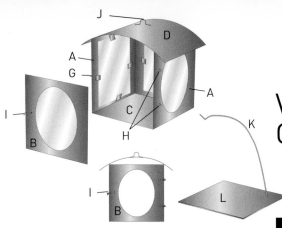

VOTIVE LANTERN WITH CHAIN AND HOLDER

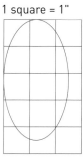

1 square = 1"

MATERIALS

- 22-gauge sheet metal
  (3, 6 × 18" sheets)
- Glass (7 × 11")
- ⅛" round rod (4')
- ¼" round rod (26")
- ³⁄₁₆" sheet metal (10 × 10")
- Hardwood stick
- 1¼" wire nails (7)

PART	NAME	DIMENSIONS	QUANTITY
A	Sides	22-gauge sheet metal × 6 × 18"	1
B	Door	22-gauge sheet metal × 6 × 6"	1
C	Base	22-gauge sheet metal × 6 × 6"	1
D	Top	22-gauge sheet metal × 9 × 6"	1
E	Glass	3½ × 5½"	4
F	Tabs	22-gauge sheet metal × ¾"*	12
G	Eyelets	1¼" wire nails	3
H	Hinges	1¼" wire nails	3
I	Hook	1¼" wire nail	1
J	Hanger	⅛" round rod × 2"	1
K	Arm	¼" round rod × 26"	1
L	Stand base	³⁄₁₆" sheet metal × 10 × 10"	1
M	Chain	⅛" round rod × 36"	1

*Cut to fit.

LAUNDRY-TUB PLANTER

MATERIALS

- ⅜" round bar (30')
- ⅛ × 1" flat bar (9')
- 10½ gal. laundry tub (1, Main St. Supply)
- ½" plywood (4 × 4 sheet)
- 2 × 4 scraps
- Wood screws
- ³⁄₁₆" round bar (10')

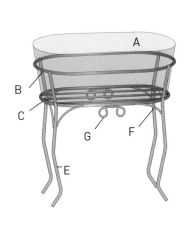

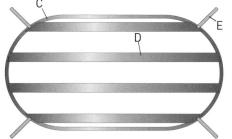

PART	NAME	DIMENSIONS	QUANTITY
A	Tub	10½ gallon	1
B	Tub support	⅜" round bar × 60"*	1
C	Base	⅜" round bar × 56"*	1
D	Crossbars	⅛ × 1" flat bar × 27"*	4
E	Legs	⅜" round bar × 25"	4
F	Leg braces	³⁄₁₆" round bar × 10"	8
G	C-scroll	³⁄₁₆" round bar × 15"	2

* Approximate dimensions, cut to fit.

THREE POST FLOOR LAMP

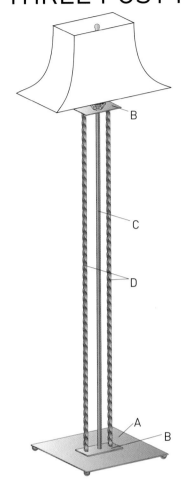

MATERIALS

- ½" round tube (48⅜")
- ½" twisted square rod (96")
- ³⁄₁₆" sheet (10 × 10")
- ³⁄₁₆ × 2" flat bar (8")
- 1" round balls (4)
- Lamp hardware and shade

PART	NAME	DIMENSIONS	QUANTITY
A	Base	³⁄₁₆" sheet 10 × 10"	1
B	End caps	³⁄₁₆" flat bar 2 × 4"	2
C	Center post	½" round tube 48⅜"	1
D	Side posts	½" twisted rod 48"	2

PATIO PLANTER STACK

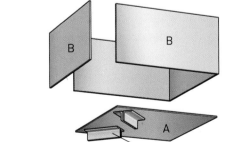

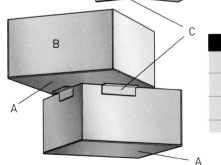

MATERIALS

- CRS sheet 14 ga. 12 × 12" (3)
- CRS sheet 14 ga. 6 × 12" (12)
- HRS steel angel iron ⅛ × 1 × 1 × 4" (8)

PART	NAME	DIMENSIONS	QUANTITY
A	CRS sheet 14 ga.	12" × 12"	3
B	CRS sheet 14 ga.	6" × 12"	12
C	HRS steel angle iron	⅛ × 1" × 1" × 4"	8

GLOSSARY

Active gas A gas, such as oxygen, that will react with other substances.

Alloy A substance composed of two or more metals or a metal and other non-metal material.

Arc welding Welding processes that use an electric arc to produce heat to melt and fuse the base metals. Often used to refer specifically to shielded metal arc welding (SMAW).

Backfire and flashback Combustion taking place inside the oxyacetylene torch creating a loud pop or explosion.

Base metal The metal that is being welded, brazed, braze welded, or cut.

Bead or weld bead The seam between workpieces that have been joined with welding.

Braze A process used to join metals using a non-ferrous material that melts above 850° F but below the melting point of the base metal. Brazing uses capillary action to join closely fitted parts.

Braze welding A process used to join metals with a filler material that melts above 840° F and below the melting point of the base metal where the filler metal is not distributed by capillary action.

Butt joint A joint between two workpieces in the same plane.

Carburizing flame An oxyfuel flame with an excess of fuel.

Consumable electrode An electrode that also serves as the filler material.

Corner joint A joint between workpieces meeting at right angles and forming an L shape.

Cutting tip Converts an oxyfuel welding torch into an oxyfuel cutting torch.

Direct current electrode negative Direct current welding where the electrode is negative and the workpiece is positive.

Direct current electrode positive Direct current welding where the electrode is positive and the workpiece is negative.

Duty cycle The amount of continuous time in a 10-minute period that an arc welder can run before it needs to cool down. Expressed as a percentage at a given amperage output.

Edge joint A joint between parallel workpieces.

Electrode The conductive element that makes the connection with the workpiece to create an electrical arc.

FCAW Flux cored arc welding.

Filler, filler metal Metal added to a welded joint.

Fillet weld A triangular shaped weld between two members that meet at 90° angles.

Flame cutting See oxyfuel cutting.

Flux A chemical compound that produces cleaning action and reduces the formation of oxides when heated.

Flux core, flux cored wire An electrode for flux cored arc welding that contains flux within a wire tube.

Flux cored arc welding (FCAW)	A semi-automatic arc welding process using an electrode wire that contains flux.
Gas metal arc welding (GMAW)	A semi-automatic arc welding process using a wire electrode, which also is the filler material. An inert gas is distributed over the weld area to shield the molten metal from oxygen. Commonly referred to as MIG (metal inert gas) or wire feed.
Gas tungsten arc welding (GTAW)	An arc welding process using a tungsten electrode, hand-held filler material, and an inert shielding gas. Also referred to as TIG (tungsten inert gas) and heliarc.
GMAW	Gas metal arc welding.
Groove weld	A weld made in grooves between workpieces.
GTAW	Gas tungsten arc welding.
Heliarc	Gas tungsten arc welding.
Inert gas	A non-reactive or non-combining gas, such as argon or helium.
Kerf	The width of a cut.
Lap joint	A joint between overlapping workpieces.
MIG	Metal inert gas—see gas metal arc welding.
Oxyacetylene cutting	Oxyfuel cutting using acetylene as the fuel gas.
Oxyacetylene welding	Oxyfuel welding using acetylene as the fuel gas.
Oxyfuel cutting	Cutting process using the combustion of a pressurized fuel gas and oxygen to heat steel to 1,600°F, at which time a pure oxygen stream is delivered to burn through the metal. Also called flame cutting.
Oxyfuel welding	Welding process that uses the heat produced by the combustion of a pressurized fuel gas and pressurized oxygen. A hand-held filler material is used. Also called gas welding.
Plasma cutting	An arc cutting process using a constricted arc. Compressed air or inert gas blows the molten metal from the kerf.
Plate	Flat metal thicker than $\frac{3}{16}$".
Saddle joint	A joint between round tubes where one tube has been cut to fit around the other.
Sheet metal	Flat metal $\frac{3}{16}$" or less in thickness.
Shielded metal arc welding (SMAW)	An arc welding process using a flux-coated consumable electrode. Also referred to as arc welding or stick welding.
Shielding gas	A gas that prevents contaminants from entering the molten weld pool.
Slag	Oxidized impurities formed as a coating over the weld bead; waste material found along the bottom edge of an oxyfuel cut.
SMAW	Shielded metal arc welding.
Spatter	Small droplets or balls of metal stuck to the base metal around the weld. Produced by shielded metal arc, gas metal arc, and flux cored arc welding.
Stick welding	Shielded metal arc welding.
T-joint	A joint between workpieces meeting at right angles and forming a T shape.
TIG	Tungsten inert gas—see gas tungsten arc welding.
Wire feed	See gas metal arc welding.

RESOURCES

Architectural Iron Designs Inc.
950 South Second Street
Plainfield, NJ 07063
800 784 7444
www.archirondesign.com

Century Trailer Inc.
868 North Bell Street
San Angelo, TX 76903
800 366 3395
www.centurytrailer.com

ChandelierParts.com
417 W. Stanton Ave.
Fergus Falls, MN 56537
218 763 7000

Decorative Iron
10600 Telephone Road
Houston, TX 77075
800 639 9063
www.decorativeiron.com

Eastwood Company
263 Shoemaker Road
Pottstown, PA 19464
800 343 9353
www.eastwood.com

e-Tarps.com
1205-202 Winter Springs Court
Louisville, KY 40243
502 584 3351
www.e-tarps.com

Garland's Inc.
2501 26th Avenue South
Minneapolis, MN 55406
800 809 7888
www.garlandsinc.com

The Lincoln Electric Company
22801 St. Clair Ave.
Cleveland, OH 44117
216 481 8100
www.lincolnelectric.com

M-Boss, Inc.
4400 willow Parkway
Cleveland, OH 44125
866 886 2677
www.mbossinc.com

Main Street Supply
PO Box 729
Monroe, NC 28111
800 624 8373
www.mainstsupply.com

Miller Electric Manufacturing Co.
1635 W. Spencer St.
PO Box 1079
Appleton, WI 54912
920 734 9821
www.millerwelds.com

National Ornamental & Miscellaneous
Metals Association (NOMMA)
PO Box 492167
Lawrenceville, GA 30049
888 516 8585
www.nomma.org

Northern Tool + Equipment
2800 Southcross Drive West
Burnsville, MN 55306
800 221 0516
www.northerntool.com

RedTrailers.com
PO Box 377
Quakertown, PA 18951
215 538 1155
www.redtrailers.com

SIC Metals & Fabrication, LLC
200 Grand Avenue
Clarion, PA 16214
www.sicmetals.com

Silent Source
58 Nonotuck Street
Northampton, MA 01062
800 583 7174
www.silentsource.com

Toll Gas & Welding Supply
3005 Niagara Lane N.
Plymouth, MN 55447
877 865 5427
www.tollgas.com

Triple-S Steel Supply Co.
6000 Jensen Drive
Houston, TX 77026
800 231 1034
www.sss-steel.com

Wagner Companies
10600 West Brown Deer Road
Milwaukee, WI 53224
888 243 6914
www.wagnercompanies.com

PHOTO CREDITS

Devilbis Industrial Finishing
195 Internationale Blvd.
Glendale Heights, IL 60139
630 237 5000
© p. 179

iStock Photo
p. 17 (top)

Lincoln / www.lincolnelectric.com
© p. 29, 30 (top), 31, 32 (top)

Miller / www.millerwelds.com
© p. 4, 8, 32 (lower)

SIC Metals & Fabrication, LLC /
www.sicmetalfab.com
p. 30 (lower) Brake design courtesy of SIC

SAFETY

Read, understand, and follow safety precautions and rules when welding. In addition to general safety rules, also reference the material safety data sheet (MSDS) as provided by manufacturers as well as the American National Standard (ANSI Z49.1) as available at these websites for free download:

American Welding Society
www.aws.org

The Lincoln Electric Company
www.lincolnelectric.com/community/safety
Welding safety publications available as PDF download (see E205 document).

INDEX